STREET, Philip. Animal migration and navigation. Scribner, 1976.
144p ill bibl index 75-30276. 8.95. ISBN 0-684-14516-2
A timely and well-chosen compilation of three decades of scientists'
interest and research in the subject field of biological migration.
Readers at all levels could profit from Street's concise yet complete
treatment. The brief bibliography could lead the more serious to the
original sources, and nearly forty maps and drawings enhance the
volume's interest and value. Recommended for general use in any
library whose clientele displays biological or bio-engineering interests.
Readable at any age level.

ANIMAL MIGRATION AND

NAVIGATION

Philip Street studied zoology at University College, London, where first he took an

honours B.Sc. in zoology and, after two post-graduate years of research, an M.Sc. Since then he has taught and lectured on zoology and natural history and written a number of popular books in these fields.

ANIMAL MIGRATION AND NAVIGATION

PHILIP STREET

CHARLES SCRIBNER'S SONS
NEW YORK

1 3 5 7 9 11 13 15 17 19 I/C 20 18 16 14 12 10 8 6 4 2

Printed in Great Britain
Library of Congress Catalog Card Number 75-30276
ISBN 0-684-14516-2

CONTENTS

page

Introduction   7

1   Migration in Fish and Other Marine Creatures   9

2   Electricity as a Navigational Device in Fish   36

3   Navigation by Sound   57

4   Migration and Orientation in Land Vertebrates   68

5   The Facts of Insect Migration   75

6   Navigation and Orientation in Insects and Other Invertebrates   97

7   Migration and Navigation in Birds   115

8   Navigation Using the Earth's Magnetic Field   137

Bibliography   141

Index   142

INTRODUCTION

The majority of animals are quite content to spend the whole of their lives in the general area where they were born or hatched. Any movements which they may make from one place to another involve only comparatively short distances, and can occur at any time of the year. An important minority, however, do undergo regular journeys which we call migrations. In some species these migration journeys cover many thousands of miles, and for each species they occur only at certain times of the year.

There are three main reasons why some animals have to migrate. Firstly, sometimes the area where they spend the larger part of the year may be too cold for them during the winter, so they migrate to a warmer climate for these few months. By contrast, for other species the temperature during the height of the summer is too high, so they seek cooler conditions at this time. The easiest way to achieve this is to migrate into the mountains, which need not involve a very long journey.

Secondly, other species have to migrate because little or no food would be available for them during the winter. This is particularly true of insect-eating birds which spend the summer and nest in temperate climates. Here the supply of insects virtually ceases with the onset of autumn, so the birds must migrate to warmer climates for the winter, where there will be an abundance of insect food for them. It is for this reason that swallows and martins fly from Europe to Africa for the winter.

The third main cause of animal migration is the need to find the conditions necessary for breeding. An area may be quite satisfactory for normal life, but unsuitable as a breeding habitat. This type of migration occurs particularly among aquatic creatures, such as salmon and fur seals.

But whether an animal is migratory or not, it has to be able to find its way about, that is to navigate. The facts of animal migration are remarkable enough, but the means by which some animals, both migratory and non-migratory, are able to navigate are well-nigh incredible. How animals achieve their journeys, long or short, we do not always know, but year by year research is throwing more light on one of the most fascinating aspects of animal behaviour. For many examples of migration we are able to offer an explanation as to how the animals navigate, or at least offer a plausible theory. But in some cases we still can do no more than record the remarkable facts and leave it to future research to provide us with an explanation.

Navigation of course involves the use of the senses, and the rapid increase in our knowledge of animal navigation which has taken place over the past few decades has been made possible by a similar rapid increase in our knowledge of the sense mechanisms with which animals are endowed.

CHAPTER 1

MIGRATION IN FISH AND OTHER MARINE CREATURES

True migration involves travel from one area at a certain time of year, and a return journey at another time. Some of the most spectacular migration journeys are undertaken by fish. In most cases we can describe the journeys but can offer little or no explanation as to how the fish manage to navigate, sometimes over vast distances.

AMAZING MIGRATIONS OF FRESHWATER EELS

Perhaps the most remarkable of all fish journeys are those undertaken by the so-called freshwater or common European eel (*Anguilla vulgaris*). Until about fifty years ago little was known about how they reproduced. The adults, known as yellow eels, were common in all European rivers from the Mediterranean to the extreme north of Scandinavia, and they were often found in lakes and land-locked ponds as well. Year by year they grew in size, until the females achieved a length of up to 3ft or more. The smaller males seldom exceeded 20in in length.

Towards the end of the summer these full-grown specimens, which were usually about six years old, suddenly changed to a silver colour, ceased feeding, and began to travel downstream until they reached the sea. By this time they had become sexually mature, the silver colour representing their breeding dress. Those living in ponds waited for a wet night, when they would leave the pond and travel overland to the nearest river. These migrating adults were never seen again, for no silver eels ever returned from the sea to the rivers. Clearly when they left the rivers they went somewhere to spawn, though there were no clues as to where the spawning grounds might be.

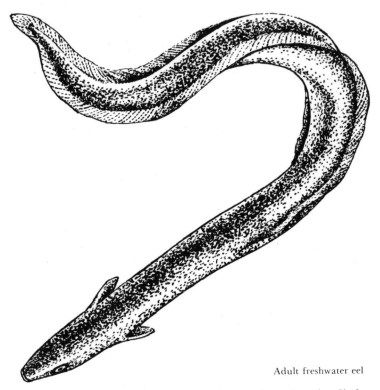

Adult freshwater eel

What was known, however, was that each spring tiny little eel-like creatures called elvers mysteriously appeared in coastal waters and proceeded to enter and travel up the European rivers, swimming vigorously against the current. Before long these tiny elvers had changed to young yellow eels. One more thing was known. In 1896 two Italian naturalists had caught a curious little creature known as *Leptocephalus brevirostris* and put it into an aquarium tank. Until then *Leptocephalus* had been regarded as a strange type of fish. It was about 3in long and very flat,

10

looking something like a transparent willow leaf. Apart from the fact of its existence nothing was really known about it; thus the naturalists decided to find out more—and they did. Much to their surprise it soon changed its shape and turned into a common elver. *Leptocephalus,* then, was not a species in its own right, but simply the larva of the common eel. It was consequently renamed and is now known as the leptocephalus larva of the eel.

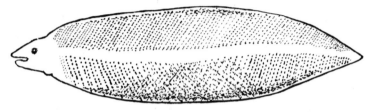

Leaf-shaped leptocephalus larva of the freshwater eel

This, then, was all that was known about the life history of the common eel up to 1920. But within a few more years a Danish naturalist, Dr Johannes Schmidt, had managed to piece together the remainder of the common eel story, and an astonishing story it proved to be. His crucial observation was that, as he travelled westward across the north Atlantic taking plankton samples, the leptocephalus larvae got smaller and smaller the nearer he approached the Sargasso Sea some 600 miles south-east of Bermuda. He was able to deduce that when the silver eels leave their European rivers they swim westwards until they reach the deepest part of the Sargasso Sea. Here they spawn, not near the surface but up to 1,000 fathoms below in a region of perpetual darkness. It is presumed that the adults die after spawning, because no spent eels have ever been found travelling back towards Europe. By marking silver eels as they left their rivers and recovering some of them by subsequent netting in the open sea it was shown that they travelled at a rate of about ten miles a day, which means that they finally arrive at the spawning grounds in the spring, having been swimming relentlessly westwards throughout the winter.

The ability of countless thousands of adult eels from the Mediterranean region right up to the farthest north of Europe to navigate across more than 2,000 miles of featureless ocean for seven or eight months and to arrive unerringly on their traditional spawning ground is indeed a remarkable feat. How do they do it? Do they possess an internal sun compass, which we now know some animals use to find their way around the world? Can they navigate at night by the stars? Or, again, are they able to use the earth's magnetic field to guide them? We just don't know. But sooner or later the biologists who are engaged on investigating the various ways in which animals navigate will be able to provide us with the answer.

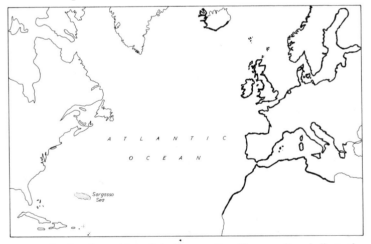

Map showing the position of the Sargasso Sea. Heavy outlines indicate the coasts from which the European freshwater eels migrate and to which the elvers return

If the westward journey of the adult eels seems remarkable, what happens to the leptocephalus larvae that are finally produced is even more astonishing. After they hatch in the depths of the Sargasso Sea they gradually swim upwards until they reach the surface waters. At this time they are scarcely ¼in long. But these tiny creatures have some built-in urge that persuades them to swim in an easterly direction, and once they set out on their incredible journey they continue to swim for a full three years, for this is the time it takes them to cross the Atlantic. They are helped on their way by the fact that they are swimming with the Gulf Stream, the steady current of water which flows continuously from the region of the Sargasso Sea towards Europe.

Of the myriads of larvae which set out a large proportion will of course never reach their destination, for they will be eaten by fish and other inhabitants of the ocean. But still a very large number do survive to reach European coastal waters, and by this time they have grown to 3in or so in length. Some time after their arrival they change shape and become elvers. After continuing to live in the sea for some months they finally swarm into the rivers the following spring. They will only leave their freshwater homes about six years later to make the return journey to the Sargasso Sea.

Three years is a long time for a larval stage to last, and it seems certain that the prolongation of the leptocephalus stage for a full three years is an adaptation to the fact that it takes all this time to cross the Atlantic. If it was not so prolonged the larvae would change to elvers in mid-ocean, and these would perish from lack of suitable food both for themselves and for the adult eels to which they would give rise.

In this connection the life history of the closely related American freshwater eel (*Anguilla restrata*) is interesting. The adults live in the rivers which flow into the sea on the eastern coasts of North America. When they reach the mature silver eel stage they swim out of the rivers and make for the Sargasso Sea, where their spawning grounds overlap those of the European eels. But from the American rivers to the spawning grounds is a relatively short journey compared with that from Europe, and the leptocephalus larvae of these American eels are able to make the return journey in one year instead of three. Correlated with this they are ready to change to elvers at the end of this year, just as they are arriving in the American coastal waters.

TRAVELS OF THE SALMON

The life history of the salmon is almost exactly the reverse of that of the freshwater eel: adult salmon spend most of their time at sea, returning to the rivers only to spawn. And whereas adult eels do all their feeding in fresh water, ceasing to feed as soon as they enter the sea to begin their spawning migration, salmon only feed in fresh water in the young stages. When, as adults, they return to the rivers to breed, they cease serious feeding as soon as they enter fresh water, though fortunately for the angler they seem unable to resist snapping at an insect which drops on to the surface in front of them as they travel upstream.

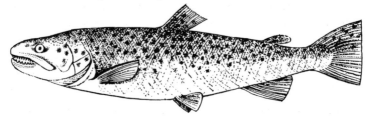

Adult male Atlantic salmon

The story begins with mature salmon coming in from the sea to breed. So strong is the driving force to reach a suitable breeding ground that they are able to mobilise terrific reserves of energy which enable them to leap over falls and weirs necessitating vertical jumps of 6ft or more. Suitable breeding grounds are found in the upper reaches of a river's tributaries, where the water is fast-flowing and clear, and where the bottom consists of clean stones or gravel with no mud.

Having achieved their goal the fish excavate shallow depressions called redds in the river bed by lying on their sides and lashing their tails to scatter the gravel. The female fish then lay their eggs in these redds while the males shed their sperm over them to fertilise them. This done, the salmon spread the gravel back over the eggs to cover them before setting off downstream to make their way back to the sea. By the time they get there they are in poor condition, not having fed for several months and having expended a great deal of energy swimming up to the breeding grounds and spawning. For many of them it will be their last spawning migration, being too exhausted to recover.

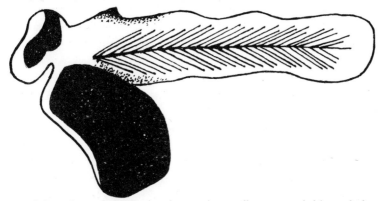

Salmon fry or alevin showing the prominent yolk sac suspended beneath the body. The yolk contains sufficient food to nourish the tiny fish for several weeks after it has hatched

Meantime the fertilised eggs are developing beneath the stones, reasonably well protected both from potential predators and from the danger of being swept downstream by the swift-running water. And this sparkling water brings them a plentiful supply of oxygen. Salmon eggs are large compared with those of many other fish, and relatively few are produced—about 800 to each pound of the female's body weight. Compared with this many other fish produce vast quantities of eggs, as we shall see. But these large salmon eggs are provided with considerable quantities of food material in the form of yolk, enabling them to indulge in a prolonged development which may take anything from five to twenty weeks. Development is slow because the eggs are usually laid during the winter. And even when they hatch the tiny salmon fry, known as alevins, still have a prominent yolk sac suspended beneath their stomachs, providing them with several more weeks' supply of food before they need to start feeding for themselves. During this early phase of development the young fish remain within the shelter of the redd.

So far what we have described is true for all species of salmon, but for the rest of the story it is necessary to distinguish between two groups of salmon, the Atlantic and the Pacific. There is only one species of Atlantic salmon, *Salmo salar,* and it is found both in European and in eastern North American rivers. Ascending the rivers of western North America, which flow into the Pacific Ocean, there are five salmon species. These are not closely related to *Salmo salar,* but are well known because they are the salmon which are canned in huge quantities and sent all over the world to be bought as 'tinned' salmon. Experiments with them have thrown much light on the problems of salmon navigation. The three most important of these Pacific salmon are the sockeye or red salmon (*Oncorhynchus nerka*), the humpback or pink salmon (*Oncorhynchus gorbuscha*) and the chinook or king salmon (*Oncorhynchus tschawytscha*).

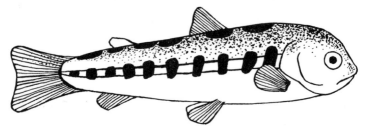

Salmon parr showing the ten prominent markings along the side of the body

But to return to *Salmo salar*. Some weeks after hatching, the alevins, now about 1in long, emerge from the shelter of

16

the redds and change to parr, which can be identified by a row of ten dark oval markings along each side of the body. Eventually salmon parr lose their characteristic markings and become silver in colour; they are now known as smolts, and it is these which now migrate down the river and swim out to sea. Some parr will change to smolts after one year, but in more northerly rivers they may live for four years before changing, the majority probably living for two years.

Once they reach the sea the smolts grow rapidly on a diet which includes herrings, sand eels and other small fish, and various shrimp-like crustaceans which occur in abundance in the plankton. The red colour of the salmon is thought to be derived from crustacean pigments. The length of time they remain at sea before returning to the rivers to spawn varies. Some return after one year, and these are known as grilse; they will certainly live to return again in later years to spawn. Others may remain in the sea for two, three or even four years before returning for the first time.

It has always been something of a mystery where the salmon live during their time at sea. Very few are ever caught by trawlers, and it therefore seems likely that they swim in the surface waters where there is an abundance of food. Here they would be out of reach of the trawl. Recent evidence has suggested that they do in fact travel much farther away from their home waters than was at one time thought, probably migrating northwards into Arctic waters which are very rich in floating crustacean life. One salmon tagged in a Scottish river was recovered eleven months later 1,700 miles away in a Greenland fjord.

Most of what we know about the way in which salmon navigate has resulted from a series of researches on the Pacific salmon species. These are more completely marine than *Salmo salar*, the young swimming down the rivers and into the sea within a few months of hatching, while they are still only a few inches long. The vast majority of these tiny fish will of course fall victim to larger fish, only a small minority surviving long enough to become adult breeding specimens ready to return to the rivers to spawn. Pacific salmon stay at sea for between four and seven years, depending upon the species, before they are ready to breed, and during this time they are known through tagging experiments to be able to travel well over 2,000 miles across the Pacific.

For a long time it had been stated without any real evidence that adult salmon returned to spawn in the rivers where they themselves had been hatched. This was finally and convincingly proved in the 1930s by a Canadian biologist, Dr W. A. Clemens. In one of the many tributaries of the Frazer river he caught nearly half a million salmon fry and tagged them before they set off down river on their journey to the Pacific. In later years, when those that survived had grown to maturity, he managed to catch no fewer than 10,958 of them in the same tributary as they returned to spawn. And despite a thorough search not a single tagged fish was found in any other tributary of the Fraser system. So not only were the returning fish able first to locate the mouth of the Fraser river itself, but they were then able to choose the correct one out of many tributaries they passed on their way upstream.

These, then, are the facts of salmon migration and navigation. And they raise two extremely interesting questions. First, how do mature salmon, perhaps thousands of miles away, manage to cross an ocean and arrive along the coastline where their particular home river enters the sea; and secondly, having correctly located the right stretch of coastline, how can they recognise the correct river, and then the correct tributary as they swim up it? To the second of these questions we can now give an answer but, for the first, we can as yet only hazard a guess based upon what is now known about similar navigation in other animals.

The first experiment designed to throw light on how salmon choose the right stream was carried out in 1957 by

Dr L. R. Donaldson and Dr G. H. Allen, two American fishery scientists. They took a large batch of salmon spawn from one west Canadian river and deposited it in the upper reaches of another river which opened to the Pacific several hundred miles away from the mouth of the first. Before the fry resulting from this spawn were ready to migrate downstream they were tagged. Several years later a careful watch was kept on the salmon returning to spawn to find out which river they would return to. And the answer was quite definite. The only marked salmon which were caught had swum into the river into which the spawn had been transported. Not a single fish returned to the river in which the spawn had been originally deposited. Clearly the fish had retained a memory of a particular kind of water, and not of a geographical location.

The result led an American biologist, A. D. Hasler, to suggest that the salmon were guided by an acute sense of smell, and that water from any particular stream had its own characteristic odour. In 1962 he proved his point conclusively. He collected over 300 salmon which had just arrived to spawn in one of the tributaries of the Issaquah river, and carried them back below the point where the tributary joined the main stream. Before returning the fish to the water he tagged them all and blocked the nostrils of half of them. On their release, those with free nostrils showed no hesitation in swimming back to the confluence and up the correct tributary. But the other half had completely lost their sense of direction, and swam about aimlessly, unable to decide where to go. Clearly the sense of smell of salmon is vastly more acute than our own, which admittedly is extremely poor. In later chapters we shall see that other animals also navigate by means of an equally acute sense of smell.

But however acute, smell could not possibly guide a fish across 2,000 miles or more of ocean. How this part of its navigation is achieved we do not as yet know. As already mentioned in connection with eels, we do know that some animals are able to use celestial navigation to guide them over great distances, and that others are able to make use of the earth's magnetic field. Whatever the answer, we are likely to know it before too long as investigation into the whole field of animal navigation proceeds.

RIVER-SPAWNING SEA LAMPREYS

Another fish which spends most of its adult life in the sea, but which swims into rivers to spawn is the sea lamprey (*Petromyzon marinus*). The lampreys and their relatives the hag-fishes comprise a curious group known as the cyclostomes or jawless fishes. Instead of upper and lower jaws provided with teeth, the cyclostome mouth is surrounded by a toothed sucker, which is used to attach the fish temporarily to a victim when it needs to feed, for these cyclostomes are parasites, rasping holes in their victims' sides and sucking their blood, rather than killing them outright.

After spending a number of years at sea the adult sea

Adult sea lamprey; (above) the toothed sucker seen from beneath

lampreys become sexually mature and ready to spawn. They now leave the sea for ever, swimming into the rivers and proceeding upstream until they come to a suitable spot for excavating a nest. This will be where the river bed is stony and free from mud. Their nests are similar to salmon redds—shallow depressions in the river bed—but the method of excavation is different. The stones are removed one by one, usually by the male, but if he comes across a stone which is too heavy for him to lift on his own his mate, who is standing by ready to spawn, may well give him a helping sucker. When the nest is ready the female lays her eggs in it while the male pours his sperm over them. The pair then replace the stones, not individually by using their suckers, but by vigorous body movements. The combined efforts of preparing and filling in the nest, and spawning, leave the adults so weak that shortly afterwards they die. So each adult lamprey is able to spawn only once during its lifetime.

Lamprey eggs are quite large, similar in size to frogs' eggs, and when they hatch they produce tiny transparent larvae about ½in long. These had been known for a long time before their connection with lampreys was established. They were thought to be a small fish in their own right, and were consequently given the name *Ammocoetes branchialis*. When their true connection was finally established they were renamed the ammocoete larvae of the sea lamprey.

Although these ammocoete larvae continue to live in the rivers where they were hatched for two or three years, they are seldom seen because they lie buried in the river bed with only their heads projecting. There is no sucker around the mouth and they are blind because, although they have eyes which will become functional when they become adult, at this stage the eyes are buried beneath the skin. From the river they take in a continuous stream of water. This provides the oxygen they need and also their food, for out of it they filter minute organisms and food particles.

Finally comes the time for metamorphosis, when the adult mouth suckers develop and the eyes come to the surface and begin to function. They are now young lampreys, ready to forsake their buried existence and swim down the river to the open sea. By this time they will have reached a length of about 6in. Over the next few years they will grow to 3ft in length before finally becoming mature, to ascend the rivers for their one and only spawning migration.

SPAWNING MIGRATIONS OF HERRING

Historically the herring has always been one of the most important branches of the fishing industry in Britain and north-west Europe. Research during this century has pieced together the full story of the regular mass migrations undertaken by herring shoals consisting of countless millions of fish. Most of our marine food fishes are demersal, living and feeding on or near the sea bed, but the herring are pelagic, swimming in the surface or middle waters and living on the multitudes of tiny plankton organisms which are found there.

The appearance of herring shoals off different parts of the British coast at various times of the year follows a well-defined pattern which, so far as we know, has not changed substantially for centuries. The herring fishery is of course closely bound up with the appearance of the shoals. Early in May shoals appear in the seas around the north of Scotland and off the Shetlands and Hebrides, and for a few months herring are being landed at the northern Scottish ports. By the time they have disappeared from these northern waters other shoals have made their appearance further south, and boats based on the eastern Scottish ports come into action, followed a little later by those from the English north-eastern ports. This continuous fishery ends with the great East Anglian season, based on Yarmouth and Lowestoft, lasting from September until well into December. In each region the herring disappear at the end of the season as suddenly as they appeared at the beginning. While the shoals are making their appearances progressively further south off the eastern shores of Britain, smaller shoals are making corresponding appearances off the west coast, culminating in a small-scale autumn herring season based on the Isle of Man.

What is the explanation of the progressive appearances and disappearances of the herring shoals? For a long time it was believed that all herring wintered in the Arctic seas, assembling in spring to swim towards the British Isles from the north-west in one gigantic shoal which divided off the north of Scotland, the larger part gradually moving down the east coast to reach East Anglia by the autumn, while a small western shoal moved southward at a similar rate as far as the Irish Sea. Certainly the times when the shoals appeared at different points along the coast fitted in with this theory. It has now been established, however, that the herring which appear in British waters belong to a number of quite distinct shoals, each with its own breeding and feeding grounds, and with its own breeding season. The quality of the herring varies from shoal to shoal, and some herring dealers claim to be able to tell where a batch of herring has come from.

For most of the year these shoals live in the middle waters well away from our shores, some probably at greater distances than others. The main reason for the annual migration which each shoal makes into coastal water is to visit its chosen breeding grounds to spawn. It is during these journeys to and from the breeding grounds that heavy catches are made.

It is not known what determines the beginning of the breeding migration of each shoal, but changes in water temperature, to which marine animals generally are very sensitive, may be an important controlling factor. In its spawning habits the herring is unique among our food fish. All other species spawn in the upper layers of the sea and the fertilised eggs, being slightly lighter than sea water, remain near the surface. Here they drift until they hatch, making a considerable contribution to the temporary plankton. Herring, however, produce eggs which are slightly heavier than sea water, and they deposit them on the sea bed in fairly shallow water, choosing ground which is stony and free from mud and weeds. After spawning the fish return to the middle and upper layers. The eggs remain in the crevices between the stones until they hatch. The eggs of other fish drifting among the plankton receive no protection, and a large proportion are eaten by fish and other marine animals before they get a chance to hatch. Among the stones the herring eggs get a certain amount of protection, though haddock are particularly fond of them. In their search for the breeding grounds of the herring fishery scientists were helped by catches of 'spawny' haddock, that is haddock which had been feeding on herring eggs and had their stomachs filled with them.

There is an interesting theory which seeks to explain the unusual breeding habits of the herring. According to this its life history was originally similar to that of the salmon, the adult fish living in the sea but coming into rivers to spawn. Here they would have had to fix their eggs to stones on the river bed to prevent them being swept downstream and out to sea. This would have been during the time that Britain was part of the continent of Europe, when the present North Sea was a plain across which the Rhine flowed northward to enter the sea north of what is now Scotland. The herring, it is suggested, would have swum into the Rhine and other rivers flowing in a similar direction at the breeding season. Then, when this great plain became inundated by the sea, the herring persisted with their traditional spawning migration, but now had to fix their eggs on the shallow sea bed.

In this connection it may be significant that two other members of the herring family, the allis shad (*Alosa alosa*), and the twaite shad (*Alosa finta*), which are almost indistinguishable from herring except for some differences in colouring, and which are often taken along with herring in the fishermen's drift nets, still do come into rivers to spawn. Outside the breeding season they generally lead independent lives, forming shoals only at the approach of the breeding season before beginning their spawning runs into the rivers.

MOVEMENTS OF SARDINES AND PILCHARDS

A fish closely related to the herring is the pilchard or sardine (*Clupea pilchardus*). It is essentially a fish of the warm waters of the Mediterranean and the Atlantic off the coast of Portugal and the Bay of Biscay. The immature fish are known as sardines, and it is in these areas that vast quantities are caught and canned. But although they live for most of their lives in warm water, the full-grown adult fish, the pilchards, come further north into colder water to breed. Their main breeding grounds are off the coast of Cornwall.

Pilchards or sardines are thus migratory fish. In the early months of the year they are confined to the warmer waters, and at this time enormous numbers of them congregate in Biscay. As we shall see in Chapter 7 these shoals play an important part in feeding the shearwaters which nest in spring on the islands off the west coast of Wales. Each year as they get older the fish migrate progressively farther north before returning south for the winter, but only the full-grown mature adults come up as far as the coast of south-west England to spawn. In earlier times pilchards from July onwards came right into the Cornish bays, at places like St Ives and Newquay, and immense numbers were caught in seine nets. Plenty of pilchards still come to their traditional spawning grounds, but they no longer swarm into the bays, and must now be caught further out to sea with drift nets.

Dr G. W. Steven of the Marine Biological Association of Great Britain has suggested an interesting explanation for the failure of the pilchards to come close inshore as they did in former times. He suggests that it is linked with the great increase in hake catches made by trawlers operating in these south-western waters. It is only in these waters that hake is really abundant, and until this century it was not considered a desirable fish. In consequence few were caught. But as its popularity increased, so too did the numbers netted. The hake is a voracious feeder, and the pilchard is one of the fish on which it feeds so that, when hake was plentiful the pilchard shoals took refuge in shallow inshore waters, where they were comparatively safe from an enemy which preferred to remain in deeper offshore conditions. As the hake population became reduced through increased fishing, so this pressure on the pilchard shoals disappeared, and they were consequently able to remain in offshore waters, which they seem to prefer, without suffering unduly heavy losses.

Two crucial questions concerned with the migration of pilchards remain as yet unanswered. Why does an essentially warm-water species need to travel north when it reaches maturity to lay its eggs in relatively cool water, and what internal urge drives the tiny sardines which hatch from these eggs to swim relentlessly south until they reach the warmer waters in which they will spend the rest of their lives until they, in their turn, become mature and ready to spawn?

MACKEREL WHICH DISAPPEAR DURING WINTER

Like the herring and the pilchard, the mackerel is a commercially important pelagic fish with an interesting pattern of migration. Until investigations were begun in the 1930s their appearances and disappearances were a mystery. What was known was that during the autumn they disappeared and that throughout the winter few were caught; but where they had gone no one knew. Then, with the approach of spring, first one shoal and then another appeared off the south-western coasts of Britain until, by April and May, large catches were being made by drifters from Newlyn and other Cornish ports.

Today the main features of the mackerel story are known. Unlike other pelagic fish, the mackerel forsake the surface waters in the autumn, sinking down to the sea bed to spend the winter in depths of from forty to 100 fathoms or more. During this demersal phase they live in gullies or along the sides of sandbanks, and not on the flat sea bed, so they are seldom caught in any quantity by the trawlers. Food is not particularly abundant on the sea bed during the winter, but the mackerel do manage to get a certain amount to eat. Each mackerel shoal is believed to have its own winter quarters to which it migrates every year. Such quarters are spread over a wide area from the English Channel out to the western edge of the continental slope 150 miles or more to the west of Land's End.

With the approach of spring the annual spawning migrations begin, each shoal becoming more active and rising to the surface, and then swimming to its spawning ground. These are not scattered over a wide area like the winter quarters, but are more concentrated, being mostly confined to two large areas well out in the Celtic Sea to the south of Ireland and westward from Land's End. Spawning is spread over five months from the beginning of March until the end of July, the earliest occurring far out to sea with subsequent spawning occurring progressively farther eastwards and therefore nearer to land.

The long spawning season is due to several factors. One of these is that the various mackerel populations do not begin their spawning journeys at the same time. Another is that some, whose winter quarters are a long way from the spawning grounds, have much greater journeys to make than others and therefore take longer to get there. Lastly, the mackerel is unusual in that its reproductive products do not all mature together, the eggs ripening a few at a time so that each individual fish takes several weeks instead of a few hours to complete its spawning.

The mackerel shoals which begin their spawning migrations in February and early March are mostly fasting, not from choice but because there is little or no plankton in the surface waters for them to feed on so early in the year. In experimental catches, however, occasional mackerel were found with their stomachs full of small fish, usually *Maurolicus pennanti*. This fish is known to move in small compact shoals, and presumably the well-fed mackerel had encountered such a shoal. Most plankton feeders—the herring for example—are unable to feed on larger organisms, but the mackerel is equipped to deal with both types of food.

Although it is not known for certain, it is believed that the earliest shoals to begin migrating to the spawning grounds are those that have wintered in the warmer Atlantic waters near the edge of the continental shelf in the Celtic Sea; and that those which have wintered in the colder waters of the Channel are the last to migrate. As more and more migrating shoals reach the spawning grounds the intensity of spawning increases, and by mid-April has reached a peak which lasts until the end of May. The intensity then falls away until the end of July, when spawning is almost at an end. In the North Sea mackerel populations, which live in considerably colder water and spawn in the Skagerrak, the maximum spawning intensity occurs a month later than it does in the Celtic Sea. Mackerel spawn near the surface, but their fertilised eggs are slightly heavier than sea water and consequently sink slowly to the sea bed.

After spawning has been completed, by early August the behaviour of mackerel changes again. The vast shoals which have characterised the previous months now split up into numerous small shoals which forsake the deeper offshore waters for shallow inshore ones, where they will remain until they begin to make once more for their winter quarters from the end of October. During these months they can no longer be landed by drifters, but can only be caught by hook and line. Their feeding habits have changed, too, and they now live mainly on young herring, sprats, sand eels and other small fish.

GRUNION WHICH SPAWN UNDER THE INFLUENCE OF THE MOON

The grunion (*Leuresthes tenuis*), is a small fish 6 or 7in long which lives in Californian coastal waters and regulates its breeding behaviour according to the phases of the moon. In order to understand its very peculiar breeding habits it is necessary to know something about how the moon affects the tides. On any coast the tide does not come up to the same point on the beach every day. At full moon and again at new moon a fortnight later the tide comes higher up the beach for a few days than it does during the periods in between. These higher tides are known as spring tides, the lower tides in between being known as neap tides.

The importance of this to understanding the breeding behaviour of the grunion lies in the fact that at the breeding season this little fish comes in from the offshore water to lay its eggs as far up the beach as possible. And so it chooses the spring-tide periods. During the four months from March to June at each spring tide a number of mature grunion swim as far up the beach as they can get. Once they have reached the very edge of the water the females raise up their bodies and wriggle into the sand tail first, the males coiling themselves round the females. When the pair have buried half their bodies in the sand the females shed their eggs and the males shed their sperm at the same time. After the fertilised eggs have been buried they continue their development, but the young fish do not emerge for a fortnight, when once again the spring-tide waters are reaching far enough up the beach for the baby fish to emerge into water and swim out to sea. If they emerged earlier they would find themselves on a dry beach, and would perish.

Investigations have shown that the eggs are probably ready to hatch a week after laying, but will only do so when they are subjected to the shaking produced by the next high tide a fortnight or so after they have been laid. This evidence has resulted from experiments on fertilised eggs kept in aquarium tanks in the laboratory. If they are shaken a week after fertilisation they will hatch, but if they are not shaken they will go well beyond the normal two-week period before hatching. The facts of grunion spawning are not disputed. But how these little fish know when the tides are high or low remains a mystery.

VIBRATION SENSE OF FISH

While we know little about the long-distance navigation of fish, we do know something about the way in which their everyday movements are modified by the information received from the lateral-line sense organs which most fish possess. If you look at a fish you will see the lateral line as a distinct line running along each side of the body from just behind the head to the tip of the tail. On the head it divides into three branches which are not so easy to see. One branch dips downwards and travels along the lower jaw, a second runs along the upper jaw below the eye, while the third runs above the eye to the tip of the snout on the top of the head.

The lateral line is really a canal running just beneath the skin and opening to the exterior by minute pores occurring at intervals along its length. It is filled with jelly-like mucus. In the canal, and also arranged at intervals, are tiny sense organs known as neuromasts. These are sensitive to changes in the pressure of the water surrounding the fish, and are therefore also capable of detecting vibrations, for vibrations produce pressure waves. Each neuromast organ consists of a

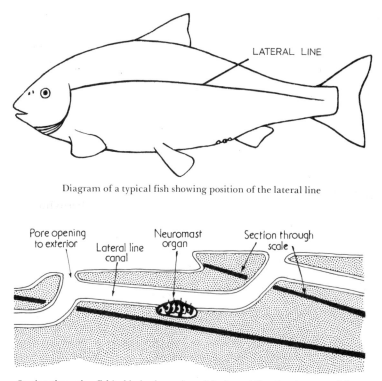

Diagram of a typical fish showing position of the lateral line

Section through a fish's skin in the region of the lateral line showing two of the pores through which the lateral line canal opens at intervals to the exterior. On the floor of the canal between these two openings a neuromast organ is shown. This consists of a number of sensory cells each having a hair-like process projecting into the canal. These hairs detect any vibrations or pressure waves entering the canal through the pores

group of sensory cells each with a single hair-like structure projecting into the canal mucus. Nerve fibres run from the sensory cells to the brain.

These nerve cells are sending back a constant stream of nerve impulses to the brain even when the fish is at rest in still water. If it swims forward in water clear of any obstructions, the pressure becomes greater on the head than along the sides, and this registers in changes in the impulses, which, as it were, report normal forward movement with no

31

obstacles ahead. But if the fish swims towards any object this will increase the pressure on the head above normal. This will be reported to the brain, and the fish will take avoiding action. This can be seen in fish swimming in an aquarium tank. As a fish approaches the glass, it is probably not able to see it, but its neuromast organs give due warning and the fish turns aside without banging its nose.

A fish can also detect any other fish swimming in the vicinity. Such swimming creates pressure vibrations which are detected by the fish's neuromast organs and thus provide it with information as to the whereabouts of the intruder, because the greatest disturbance will be caused to the neuromasts which are closest to it. A fish swimming in daylight in clear water will probably gain much of its information about its surroundings through its eyes, but in very muddy water and at night it probably relies mainly upon its lateral-line system. To fish which are adapted to live in perpetual darkness in caves the system must be invaluable. In fact the whole lateral-line system is particularly well developed in all cave fish.

BREEDING MIGRATIONS OF SEALS

Although mammals are primarily terrestrial animals, two groups have successfully abandoned life on land for an aquatic existence at sea. These are the Cetacea, comprising whales, dolphins and porpoises; and the Pinnepedia, comprising seals, sea-lions and walruses. The Cetacea are fully aquatic, even mating and giving birth to their young while still in the water, but the Pinnepedia have to abandon the water for a few weeks every year in order to give birth to their young and to mate almost immediately afterwards. Seal pups have to grow up fast, because within a few weeks of birth they are abandoned by their parents and have to take to the water to fend for themselves. To enable them to treble their birth weight in this short time the milk which their mothers provide for them is incredibly rich, containing in most species more than 50 per cent of fat and a high concentration of protein.

An important feature of the seals' breeding behaviour is that for any species there is only a limited number of breeding grounds, so that on each ground a very large number haul out at the breeding season. In some cases a million or more adults may return to the beaches of quite a small island. But when they put to sea again they disperse far and wide as individuals, and may travel 1,000 miles or more during the ensuing year. Yet when the next breeding season approaches they all manage to find their way back to their own breeding grounds. The pups in fact will stay away until they themselves become mature, usually at three years old, and still find their way back to the place where they were born.

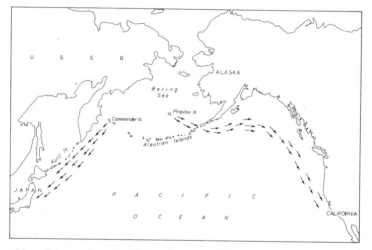

Map of the northern Pacific showing the positions of the Commander and Pribilov Islands and the migration routes of the fur-seal populations of the two island groups

The most spectacular seal migrations are those undertaken by the Northern fur seals (*Callorhinus ursinus*). Their breeding grounds are two groups of small islands, the Pribilov Islands and the Commander Islands, situated far north in the Bering Sea between Alaska and Siberia. Once the breeding season is over the Commander Island seals swim south-westward into the Pacific, many of them reaching Japanese waters, about 3,000 miles away. Those from the Pribilov Islands swim south-eastward to reach the coasts of California, covering a similar distance. Those individuals which reach the southernmost limits thus achieve a round trip of some 6,000 miles in eleven months or less. To be able to home accurately on a small group of islands from such immense distances represents an incredible feat of navigation. The facts of these astonishing journeys are beyond dispute, but once again we must wait for the results of future investigations before we know how they are accomplished.

REGULAR BREEDING AND FEEDING MIGRATIONS OF WHALES

Unlike the seals, whales have become completely adapted to an aquatic life, mating, birth and suckling all occurring in the water. But most species undergo annual migrations which may take them over distances of several thousand miles. The reasons for these migrations are that, like the fur seals, their breeding grounds are a long way from their feeding grounds.

There are two kinds of whales: the toothed whales, comprising sperm whales, killer whales, dolphins and porpoises; and the baleen or whalebone whales, which despite their huge size feed on small plankton organisms which they filter out from the sea water by means of huge plates of baleen which hang down from the roof of the mouth. These large whalebone whales form two distinct groups, one group being confined to the southern hemisphere and the other to the northern. The most important southern species are the blue whale, the fin whale and the humpback whale. Incredible as it may seem, these giant whales feed exclusively on a single species of shrimp-like crustacean an inch or two in length, the euphausiid *Euphausia superba,* commonly known as krill. The number consumed by a full-grown blue whale, which may weigh anything up to 130 tons, must be astronomical. Krill, however, do not occur in temperate or subtropical regions, only in the cold Antarctic waters at the edges of the icefields. And it is here that the whales spend the summer, eating enough not only to satisfy their present needs, but to provide a store to carry them through the following winter.

As the Antarctic winter approaches, so the whales swim northwards to their breeding grounds in the subtropical regions of the south Pacific, south Atlantic and Indian oceans. Here mating takes place, and those females which mated the previous year give birth to and nurse their calves. In these warmer waters they spend the winter, but there is nothing for them to eat, there being no krill, and apparently they will not eat anything else. By the spring the young calves will be well grown, and when they accompany their mothers as they return to the Antarctic they are able to begin feeding for themselves. The gestation period is ten to eleven months, and since birth is followed by about six months nursing, the females only breed every other year.

In the northern hemisphere the story is similar. During the summer the whales are feeding in the far north. But these northern species are not so exclusive in their diet, taking copepod crustaceans and young herring, as well as euphausiid *Meganyctiphanes norvegica,* known as the northern krill. As winter approaches the Atlantic species move southwards to their breeding grounds around the Azores and off the coast of north-west Africa. The Pacific species also migrate southwards to the central Pacific, and some may even travel as far as the Indian Ocean to breed.

How whales achieve these long-distance migrations we do not know, but recent investigations show them able to navigate over short distances, like bats, by means of echo location (see Chapter 3).

CHAPTER 2
ELECTRICITY AS A
NAVIGATIONAL DEVICE IN FISH

It has been known for a very long time that a few very specialised fish are able to produce high-voltage electrical discharges which they use to stun their prey. The African electric catfish (*Malapterurus electricus*) was known to the ancient Egyptians nearly 5,000 years ago and was figured on the walls of their tombs, and the discharges from torpedo-rays were used by the Romans as a cure for gout. About 200 years ago the South American electric eel (*Electrophorus electricus*) was discovered, and proved to be the champion of them all, capable of producing a shock of more than 500V, against the catfish's maximum of 350V and the torpedo-ray's 220V. Until recently this was about the sum total of knowledge concerning the production of electricity by fish.

In 1958, however, Dr H. Lissmann of Cambridge University published an extremely important paper entitled 'On the Function and Evolution of Electric Organs in Fish'. Since the end of World War II he had been experimenting with a certain group of African freshwater fish, and had discovered that many species were able to produce minute electrical discharges into the surrounding water. This in itself was an important discovery, but even more important was Dr Lissmann's further discovery which established the fact that the fish used their electrical abilities as means of navigation.

The discovery that large numbers of fish belonging to several quite unrelated groups in different parts of the world produced low-voltage discharges only became possible as a result of wartime research. During World War II the electronics industry concentrated on producing amplifiers and hydrophones vastly more sensitive than anything which was available before the war began,, their purpose being submarine detection. Dr Lissmann's work began on a visit he made to Ghana just after the war in order to study the wild life of the rivers, and he took with him some of these new amplifiers. Like so many African rivers, the ones he chose to investigate were extremely muddy, with so much silt suspended in the water that visibility was very severely restricted. Muddy the water might be, but the submerged amplifiers revealed an abundance of life — or at least an abundance of electrical activity. The amplifiers could be connected either to an oscillograph, in which case the discharges registered on a screen, or to a pair of extremely sensitive headphones, in which they registered as clicks and crackles.

PIONEER WORK WITH GYMNARCHUS

Dr Lissmann was able to identify five distinct kinds of discharge. One kind consisted of just a few pulses per second, another of between 20 and 50 per second. A third consisted of a very steady hum of about 300 per second, and it was on this that he decided to concentrate his attention, the aim being to identify the species which was producing this particular discharge. Soon the possessor of this hitherto unsuspected electrical talent was found to be *Gymnarchus niloticus*, a species common and well known in many muddy African rivers, including the Nile, as its name suggests.

Gymnarchus is in many respects an unusual fish, but Dr Lissmann's discoveries as a result of several years of experiments have shown the significance of its various peculiarities. It is a fairly large fish with a full-grown length of 3ft or more. Its eyes are poorly developed, and would be of little use at the best of times; in a muddy river they must be quite useless. And yet it is capable of chasing and capturing fast-moving fish. It also possesses an apparently uncanny ability to swim backwards into quite narrow crevices in the river bank without colliding with the sides.

Whereas most fish swim by means of a side-to-side motion of a powerful muscular tail, *Gymnarchus* keeps its body quite straight. It is moved by undulations of a much elongated dorsal fin, which runs almost the whole length of the body. Reversing the direction of the undulations enables it to swim backwards. The only other fins it possesses are its pair of pectoral fins. The pelvic, anal and caudal or tail fins have disappeared, and its body tapers almost to a point at the hind end.

Dr Lissmann's first clue as to the way in which *Gymnarchus* makes use of its ability to produce electrical pulses was gained as he drifted slowly in a boat on a river and held a magnet over the side a few inches above the water and soon attracted a specimen to the surface. As he moved the magnet forwards and backwards, the fish followed it. This initial experiment was followed by many others carried out in an aquarium tank. On the outside of the tank he placed a plywood board, and moved a bar magnet about outside the board, where it could not be seen by the fish. But the fish knew it was there, and followed its every movement. Similar screens of other materials, including slate and marble, were used, with similar results. Clearly the fish was able to detect and respond to the magnetic field produced by the magnet. A metal screen, however, cut out any response. It would also prevent the magnetic field passing through to the other side.

From his various experiments Dr Lissmann was able to formulate his theory to explain how *Gymnarchus* was able to use its electrical discharges both to navigate and to detect and capture its prey. If a piece of paper is placed over an ordinary bar magnet resting on a table and iron filings are sprinkled on the paper above the magnet and the table is

38

gently tapped, the filings arrange themselves in a pattern indicating the magnetic lines of force radiating from the magnet and constituting its magnetic field. *Gymnarchus*, Dr Lissmann argued, produced a similar electrical field made up of lines of force radiating from its body. Any fish or other object coming within range of this field would cause distortion of the lines of force, and these distortions would be detected by *Gymnarchus*. In this way it would be given information about any object around it.

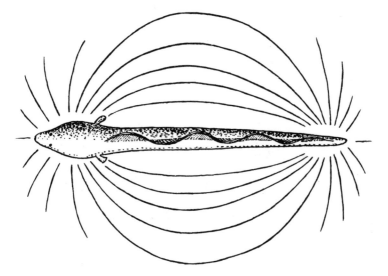

The electrical field surrounding *Gymnarchus*

This electrical field is probably also the means by which *Gymnarchus* individuals communicate with one another. In another series of experiments Dr Lissmann and his colleague Dr K. E. Machin fixed six pairs of electrodes at various points on the sides of the tank, with wires from these electrodes passing through the glass to the outside. They then recorded the discharges made by another *Gymnarchus* specimen and played these back through the various electrodes, using one at a time. Whichever electrode was used it was immediately attacked by the fish in the tank. The results of this experiment suggested that in the wild the discharges of one fish serve to warn another not to encroach upon its territory. This fits in with the fact that *Gymnarchus* specimens are very quarrelsome among themselves, so that two or more cannot be kept in the same tank. *Gymnarchus* of course can not only detect members of its own kind, but members of other species on which it preys as well. But these potential victims are not only in the visual dark, because owing to the very murky water they are not able to see their enemy approaching, they are also as it were in electrical darkness because, lacking electric organs and electric detectors, they are unable to detect the constant electrical field of their pursuers.

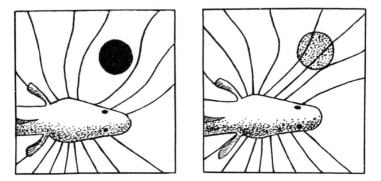

Diagram to show that a bad conductor (left) distorts the electrical field more than a good conductor (right)

Thus *Gymnarchus* 'sees' what is going on around it using an electrical field instead of light rays. Its ability to back into a crevice scarcely wider than its body can now be understood. Both sides of the crevice are detected, and it is easy for the fish to steer between the two. Also, since the field completely surrounds the body, it is just as easy for it to steer backwards as forwards and, as we have seen, *Gymnarchus* can move with equal ease in either direction. An animal which is guided by sight can steer much more accurately forwards than backwards, because its eyes are at the front of its body.

We can now understand why *Gymnarchus* keeps its body straight all the time. Any bending would distort the electrical field, and information derived from it would be inaccurate.

ELECTRIC ORGANS

Parallel with these investigations into the behaviour and ability of electric fish there has been a great deal of research into the structure of the organs which produce the electricity. In different kinds of electric fish the organs are found in different parts of the body, but they are all similar in fundamental structure. Each electric organ consists of a considerable number of modified muscle cells, each with its own nerve supply.

In *Gymnarchus* there are eight horizontal tubular organs running the whole length of the tail, four on each side of the mid-line. Each organ consists of between 150 and 200 electroplates lying vertically and packed close together like a pile of coins laid on its side. Altogether each fish possesses between 1,200 and 1,600 electroplates, each of which represents a modified muscle cell which has retained its nerve supply. The front surface is flat and the end of the nerve fibre is attached to it, but the hind surface is deeply folded. Each electroplate is embedded in a gelatinous matrix, and as the plates themselves are fairly transparent the whole organ has a clear jelly-like appearance.

AMERICAN KNIFE-FISH

Gymnarchus is but one of a group of about 150 species known as mormyroid fish which live in the muddy African rivers. They all possess electric organs, but in all the others there are only two organs on each side of the tail, containing between them 600 to 800 electroplates. Mormyroid fish are not found in any part of the world except Africa. But in

Central and South America there is another group of fish known as knife-fish which also have electrical ability remarkably similar to that of mormyroids. The two groups, however, are not related, and they cannot possibly have inherited their electric organs from a common ancestor. These must have been evolved separately in the two groups.

The fact that these knife-fish produced electricity was discovered by Dr Lissmann working with *Gymnotus carapo*, after his pioneer discoveries with *Gymnarchus*. He found that *Gymnotus* could also detect a magnet and would follow it if it was moved about. It also kept its body quite rigid and could swim forwards or backwards with equal ease, and in swimming backwards could pass quite fast through a small opening without touching the sides.

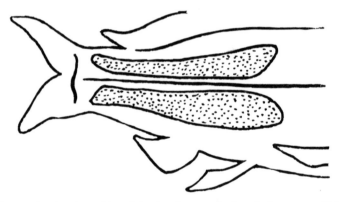

Diagram showing the position of the electric organs in the tail of a mormyroid fish

Subsequent to his work American zoologists have studied many more knife-fish, and every one so far examined has been found to produce electricity. Probably every member of the group will eventually prove to be electrogenic. Like the mormyroids, the knife-fish are propelled by the undulations of a single fin, but in this case it is the anal fin, which is extended along most of the ventral surface of the body. The dorsal and tail fins are suppressed. Their electric organs are very similar to those of the mormyroids, but they extend from the tail into the hind part of the body.

From the present state of our knowledge it seems likely that each mormyroid and each knife-fish species produces its own particular pattern of electrical discharges and therefore its own specific electrical field. With the knife-fish there seems to be some correlation between the frequency of the impulses and the conditions under which the species live. Those living in slow-moving rivers produce only a small number of impulses per second — one species being credited with only 2 — whereas those living in rapids may have an impulse rate as high as 1,600.

MORMYROMASTS

One thing we have not yet considered is how these electric fish are able to detect the changes in their electrical field brought about by the proximity either of another fish or of an inanimate object. Obviously if electrical navigation is to work the fish must not only generate its electrical field, but possess sense organs capable of monitoring changes that occur in it.

An electric sense organ or mormyromast

In fact electric fish do have electric sense organs, which are called mormyromasts because they were first identified in the mormyroid fish. Each sense organ consists of a microscopic skin pore leading into a canal filled with jelly. At its inner end the canal widens to form a flask-shaped cavity, on the floor of which there are a number of sensory cells, each provided with a nerve fibre which travels to the brain, where it ends in the cerebellum, which in electric fishes is relatively larger than it is in other species. The cells are sensitive to electricity, just as the cells of the retina of the eye are sensitive to light rays, and those of the inner ear to sound waves.

ELECTRIC EELS

The remarkable results revealed by studies of the African mormyroids and the South American knife-fish led to a renewed interest in the electrical activities of the South American electric eel, which incidentally is not an eel at all, but a relative of the knife-fish. Like them it is found in many Central and South American rivers where the water is often very muddy. It also keeps its body quite rigid and propels itself forwards or backwards by undulations of an enormously developed anal fin, which extends practically the whole length of the ventral surface. Dorsal and caudal fins are not developed. The results of this modern look at the electric eel proved most interesting. It possesses not one, but three kinds of electric organs. The main organ, which produces high-voltage discharges up to 650V with a current of up to 1A, extends for most of the length of the tail. But at the hind end of the tail it gives way to a much smaller electric organ, the organ of Sachs. Beneath both organs runs a third, the organ of Hunter.

Studies of the electric eel carried out by scientists at the New York Aquarium showed that, when it was resting, the electric eel showed no electrical activity at all. As soon as it started to move about, however, it produced small electrical

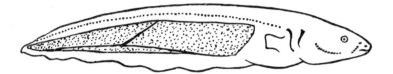

Position of the electric organs in the electric eel

discharges similar to those produced by the mormyroids and the knife-fishes. It seems likely that they serve the same purpose, providing the fish with information about its surroundings and enabling it to navigate. These discharges were shown to be produced by the organ of Sachs, which is thus comparable with the electric organs of the other two groups of fish.

The discharges of the main organ are used to stun the eel's prey. It seems likely that the electric eel regulates these discharges so that they do not actually kill. If this is so, it must be able to modify the strength of discharge according to the size of the prey. Aquarium specimens will not eat fish they have killed with too strong a discharge. They will only eat those they have stunned.

An interesting suggestion has been put forward to explain the function of the organ of Hunter. It is known that many non-electric fish can be attracted to electrodes placed in a river, and can thus be caught. The suggestion is that the organ of Hunter produces a discharge sufficient to attract fish towards the electric eel, and that when they have come close enough they are stunned by a high-voltage discharge from the main organ.

The structure of the electric eel's electric organs is similar to that already described for *Gymnarchus* and *Gymnotus*. They are however relatively enormous, accounting for about half the total body weight. The high-voltage organ consists of about 70 columns of electroplates on each side, each column consisting of between 6,000 and 10,000 electroplates. It seems likely that most or all of the electric eel's mormyromasts are situated in the head region, because when the head of a specimen was covered with a thin layer of lacquer it was no longer able to detect prey, although it was still producing a normal electrical field.

Electric eels are large fish, full-grown specimens achieving a length of 8ft or more. Young fish up to about 1ft in length have well-developed efficient eyes, but after this there is a steady degeneration and eventually they become completely blind.

To sum up, so far as we know at present, freshwater electric fish are only found in Africa and in Central and South America. There are none in North America, Europe, Asia or Australasia, though of course it is possible that further investigation may discover hitherto unknown electric fish in some of these areas. The parallel between the electric fish of Africa and Central and South America is interesting. The African mormyroids have the knife-fish as their American counterparts, and although the two groups are unrelated their electrical abilities are remarkably similar, as we have seen.

AFRICAN ELECTRIC CATFISH

There is a similar parallel between the South American electric eel and the African electric catfish. Like the eel it is a large fish, adult specimens growing to a length of about 4ft, and it produces a high-voltage discharge which it uses

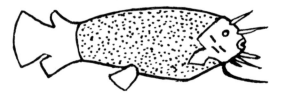

Position of the electric organs in the electric catfish

to stun its prey. Whether it can also produce a low-voltage field for navigation is not yet known, but it seems likely that it will be found to do so. It certainly lives in muddy waters, and its eyes are not very well developed. Its electric organs are situated farther forward than they are in the other three groups, and lie beneath the skin of the body and the front part of the tail.

STARGAZERS OF THE SEA BED

So far we have been discussing freshwater electric fish. Stargazers belong to one of the three groups of marine fish which are electrogenic, the others being torpedo-rays and skates. There are many gaps in our knowledge of the stargazers. They are a small group of fish which are widely distributed on the sea bed on the Atlantic and Pacific coasts of North and South America. Like the freshwater groups they are bony fish, full grown they seldom exceed 1ft in length, and their name refers to the fact that their relatively small eyes are set on the top of the head facing upwards.

Whereas the electric organs of all the electrogenic groups so far considered are derived from the muscles of the tail or trunk region, the stargazer's electric organs are situated just behind the eyes, and are in fact developed from modified eye muscles. How the electricity they produce is used is at present unknown, but the charges are sufficient to be felt by a hand put into the water in which a stargazer is living, which suggests that they may be used to stun prey.

TORPEDO-RAYS AND SKATES

Torpedo-rays and skates belong to the large group of cartilaginous fish, which also includes sharks and dogfish. They are flat fish, but they are flattened from above, unlike the bony flat fish — plaice, sole, halibut, turbot etc — whose bodies are flattened from side to side. The large 'wings' of the skates and rays, which are the parts we eat, represent enormously enlarged pectoral fins. Fish normally have two sets of paired fins, the pectoral fins which are comparable with our arms, and the pelvic fins which correspond to our legs.

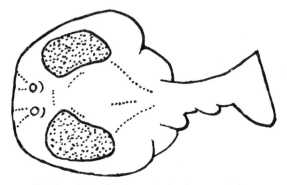

Position of the electric organs in the torpedo-ray

The electric organs of the torpedo-ray are formed from the muscles of the wings, and not from those of the body or the tail as they are in all other electrogenic fish. Each is a large kidney shaped organ occupying a considerable proportion of the wing. The arrangement of the electroplates, too, is different, each organ consisting of a number of vertical columns, with each column containing up to 1,000 electroplates. In a large specimen the total number in the two organs can approach half a million. The largest torpedo-rays may attain a length of 6ft, and these can produce a maximum shock of about 220V.

Skates and rays are generally rather sluggish fish which live on the sea bed, the majority feeding on the molluscs, crustaceans and other small animals which also live there. The torpedo-rays, of which there are a number of different species, are even more sluggish than most other members of the group. Yet in their stomachs the remains of many different kinds of active fast-swimming fish have been found, including salmon, eel, red mullet, plaice and dogfish. They could not conceivably have caught these fish unaided, and we must assume that, like the electric eel and the electric catfish, they are able to stun fast-moving fish which swim close enough to them by a burst of high-voltage discharges. Whether torpedo-rays are also able to produce a low-voltage navigational field we do not know, but at least one species has a separate small electric organ situated behind each of the main organs. Further investigation may well show these to be low-voltage organs.

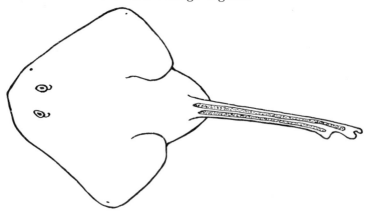

Position of the electric organs in the skate

At least some species of skate possess electric organs, but despite their relationship to the torpedo-rays their organs are quite different. They occur in the tail, and represent modified tail muscles, and the voltage they produce is quite low and certainly insufficient to stun prey. The highest voltage recorded is 4V, produced by the common thornback ray (*Raia clavata*). How these skates make use of their electrogenic ability we do not know. One constant feature of all electric fish which use their electrical field for navigation is that they keep their bodies straight. The skate, however, uses its tail to propel itself through the water with a whip-like action, which would mean that the electric field would be continually distorted, and unlikely to be of any navigational use.

FISHING BY ELECTRICITY?

The fact that certain kinds of fish use electricity to capture their prey led to the suggestion that man might also use electricity to catch fish. It was discovered that if two metal electrodes connected to a direct current supply are placed in fresh water, some of the fish in the vicinity are attracted to the positive electrode. They seem to be hypnotised by the current, and are apparently unable to break away from its influence. It is thus easy to gather them up into a net. The method has been used to catch up the fish in trout streams as a method of reducing the numbers of coarse fish. The trout are returned to the river after the sweep, but not the coarse fish.

Investigations are under way to see whether the use of electricity in sea fishing might be feasible. If it worked, fishing by electricity might have a very important advantage over trawling, because it seems likely that it might be possible to control the size of fish attracted to the electrodes by varying the strength of the current used. This would mean that only mature fish need be caught, and that small immature fish would not be put at risk. It is true that small fish that come up in the trawl are eventually returned to the sea, but a high proportion of them are seriously damaged or killed in the process.

CHAPTER 3

NAVIGATION BY SOUND

Hardly less spectacular than Dr Lissmann's discovery that certain fish used electricity as a means of navigation was the somewhat earlier discovery that bats used sound for the same purpose. This discovery was also only made possible by the development of suitably sensitive instruments. Long before this, however, scientists had been very interested in the remarkable ability of bats to fly in total darkness without colliding with objects, and to capture flying insects with ease.

DISCOVERY OF ECHO LOCATION IN BATS

The first experiments in this field were performed in 1793 by an Italian naturalist, Lazaro Spallanzani. Having caught a number of bats in a nearby belfry he proceeded to put out their eyes and release them in a room across which tapes had been stretched in all directions. The bats were not in the least inconvenienced by their loss of sight, for they were able to fly about the room among the tapes without once colliding with them or with the walls. In the following year the French naturalist Charles Jurine carried out further experiments. He plugged the bats' ears with wax, and found that they lost their ability to navigate. They fluttered around helplessly, bumping into the objects in the room as well as into the walls, seeming totally unable to negotiate any obstacles in their path.

For well over 100 years after this no further experiments were performed, but various theories were put forward in attempts to explain the bats' extraordinary navigational abilities. Spallanzani himself came nearer to the truth than anyone else when he suggested that they used echoes from the noise produced by their flapping wings to guide them. Others suggested that the bats might be sensitive to the slight increase in the air pressure in front of them as they approached a solid object. Fish, as we have seen in Chapter 1, do use this method through their lateral-line system.

It was not until 1920 that scientific attention was next directed to discovering the means by which bats were able to fly with such accuracy in total darkness. In that year a Cambridge physiologist, H. Hartridge, became interested in the bats which used to fly into his rooms through the open windows at night chasing moths. He noticed that if he switched off the light this had no effect on their ability to navigate. If a door was ajar they would fly into the next room, unless there was an insufficient gap, in which case they would not attempt the passage. Having satisfied himself that sight played no part in their navigation, he suggested that they probably acquainted themselves with the whereabouts of solid objects by producing sounds which were reflected back from them. These sounds, he suggested, were produced at a high frequency, well above the limits of human audibility. These became known as ultrasonic sounds, and the method of using them in navigation as echo location or sonar. Hartridge, however, was busy with other work, and did not continue his bat investigations although, as we now know, he had hit upon the correct explanation of the bats' navigation. Incidentally, despite the common phrase 'as blind as a bat', bats are not blind, the eyes in most species being reasonably well developed.

In the early 1920s a method of measuring the depth of water beneath a ship using echo location was developed. At first the apparatus was somewhat primitive, consisting of a hammer which struck the inside of the hull, and a timer which measured the time that elapsed between the hammer blow and the return of the echo from the sea bed. Knowing the speed with which sound travelled through water it was possible to estimate how far the sea bed was below the ship. Within a few years various improvements produced a more sophisticated and more accurate echo sounder. Today it is used by fishing vessels not only to locate shoals of fish and report on their depth, but also to identify the species to which they belong.

Much earlier than this the skippers of steamers crossing the north Atlantic in latitudes where icebergs might be expected used a simple kind of echo location to detect them in conditions of poor visibility. Periodically they would sound a short blast on the ship's siren and listen for an echo. If one occurred, it indicated the presence of an iceberg, the time lapse between the siren blast and the return of the echo giving an indication of how far away it was. Sometimes the effect was disastrous, because almost at the moment the siren blast ceased the ship ploughed straight into an iceberg and was wrecked. The explanation of these occasional disasters was that when the siren sounded the iceberg was so close that the echo returned to the ship while the siren was still sounding, and was therefore not detected, being weaker than the sound of the siren. As we shall see, this is a problem that bats have had to solve.

The fact that bats navigate using ultrasonic frequencies was finally demonstrated by Donald Griffin, an American zoologist, in 1938. He first carried out a series of experiments to establish exactly what bats could and could not do. Blindfolding caused them no inconvenience, but covering or plugging their ears as well made it impossible for them to fly among objects without colliding with them. With only one ear covered they were able to fly reasonably well, but some collisions did occur. Covering the mouth instead of the ears produced a similar result. These results fitted in with the theory that bats produced ultrasonic sounds which passed through their open mouths, and detected the returning echoes with their ears.

The problem now was to demonstrate that in fact bats did produce ultrasonic frequencies which we could not detect with our ears. Fortunately for Griffin, Professor Pierce, an American physicist, had recently developed an electronic apparatus which converted high-frequency ultrasonic vibrations into relatively low-frequency vibrations which could be heard by the experimenter. By measuring the frequency of the audible sound, and knowing the reduction factor given by the apparatus, it was possible to calculate the frequency of the original ultrasonic vibrations.

Griffin arranged to take a cage of bats to Professor Pierce's laboratory. Placed in front of the apparatus the apparently silent bats caused it to pour forth a noisy medley of sound. Professor Pierce's design was soon modified to produce a bat detector, with which Griffin and Robert Galambos, a physiologist who had joined him, were able, over a number of years, to make a thorough investigation into the whole field of echo location in bats.

Sound is produced by air vibrations, the pitch of the sound depending upon the number of vibrations or cycles per second (cps), the greater the number of cycles the higher the note. Middle C on any musical instrument produces 256cps, while the top note on the piccolo produces about 4,700cps. The limit for our ears is about 20,000cps, the figure being higher when we are young. Bats when they are flying, and often when they are at rest in caves and belfries, do produce some squeaks at about this figure. These are the squeaks which many of us could hear when we were young but can no longer detect when we grow older. These, however, are not the all-important frequencies used in echo location, which are much higher, varying between 30,000 and 100,000cps.

Using their new bat detector Griffin and Galambos first established the remarkable accuracy of the bats' sonar. A room was hung with vertical wires about 12in apart. If these were only 1mm in diameter the bats did sometimes collide with them, though they were able to avoid them four times as often as they hit them. In a series of experiments using ever thinner wires it was found that the thinner the wires the more often did the bats collide with them. Only when the diameter of the wires was reduced to $\frac{7}{100}$ mm, making them similar in thickness to a human hair, were the bats completely incapable of detecting them. Even when loud recordings of the ultrasonic sounds of other bats were played in the room, they were still able to detect the echoes from and avoid wires with a diameter of $\frac{3}{10}$ mm.

Bats face the same problem as the ship sounding its siren when close to an iceberg. In consequence the sound is produced in a series of ultrasonic squeaks of incredibly short duration, just a few thousandths of a second, and while each squeak is being uttered the ear is temporarily disconnected by the contraction of a tiny muscle. This ensures that a bat cannot hear its own squeaks, only their echoes when these return to it. And the timing is such that each echo has returned and been registered by the ear before the next squeak is uttered. The rate at which the squeaks are produced varies considerably. When the bat is flying well away from any inanimate object or any flying insect it will utter only four or five squeaks every second. These we can regard as exploratory squeaks, and they will not result in returning echoes. As soon as an echo is received, however, indicating that the bat is approaching some object, the rate of squeaking is speeded up, and continues to increase as the bat comes within a few yards of the object which is causing the echo. When it closes to within a yard or so the rate may be as high as 200 per second, which means that the bat must also be able to disconnect and connect its ears at a similar rate. As soon as the insect has been caught or the inanimate object passed the bat drops back to the normal slow exploring speed.

At the same time as the bat flies with its mouth open, it turns its head from side to side, thus exploring as wide an angle ahead as possible. Each pulse or squeak, too, is not just a short burst of sound on a given frequency, but is subject to frequency modulation. At the beginning the frequency rate may be as high as 100,000cps, but at the end, a few thousandths of a second later, it may have dropped to 30,000cps or even less.

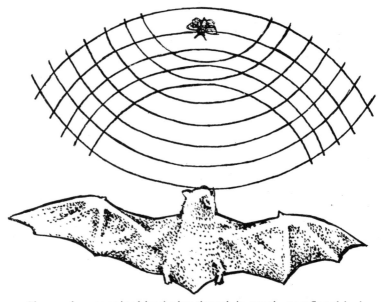

The sound waves emitted by the bat through its mouth are reflected back from the insect and detected by the bat's ears

Altogether there are about 900 different species of bats in the world, and so far the sonar of only a small minority of these has been investigated with the bat detector. But these few can be identified by their voices, since each has its own characteristic ultrasonic voice pattern. It seems likely therefore that eventually all species of bats will be capable of being recognised by their voices, given a sufficiently sensitive bat detector. This also suggests that bats themselves can probably recognise one another by their voice characteristics. This must be particularly important to them at the breeding season in recognising their mates, and subsequently their offspring.

HORSESHOE BATS AND THE DOPPLER EFFECT

The type of echo location so far described applies to the majority of bats, but there is one group, the leaf-nosed or horseshoe bats, whose sonar is substantially different. These bats have a rather grotesque appearance, the nose being surrounded by a horseshoe-shaped flap of skin known as the nose-leaf. Unknown to one another, at the same time as Griffin and Galambos were investigating the sonar of normal insect-eating bats in America, F. P. Möhres was experimenting with the ultrasonic abilities of horseshoe bats in Germany.

Head of a horseshoe bat showing the nose-leaf surrounding the nose

He was able to show that these bats fly with their mouths closed and that the ultrasonic squeaks are transmitted through the nose, and directed by the nose-leaves acting as a reflector, in the same way as the parabolic mirror at the back of a searchlight concentrates the light into a narrow beam. During flight the head is not turned from side to side as it is in other bats. Instead scanning is done by left and right movements of the nose-leaves, resulting in a continual change in the direction of the ultrasonic beam. The ears, too, are moved forwards and backwards up to sixty times a second in order to receive the echoes coming back from different directions.

Instead of occupying only a few thousandths of a second each pulse or squeak continues for about $\frac{1}{10}$ sec, with a correspondingly smaller number of squeaks per second, usually between four and six. The squeaks, too, are not frequency modulated, but consist of a constant very high-frequency in the 80,000 to 100,000cps range, the actual frequency varying according to the species.

In some way not yet completely understood horseshoe bats make use of the Doppler effect. This is the effect we hear when a train is approaching and sounding its whistle or siren. As it gets nearer so the pitch of the sound appears to rise and the note seems to get higher, but from the moment the train passes the note seems to fall, although the actual pitch produced by the whistle is on a constant frequency. When the bat is flying in the direction of an insect which is flying away from it the pitch of the echo when it reaches the bat's ears will appear to be lower than that of an echo from a stationary object, while the echo from an insect flying towards the bat will seem to be higher. Clearly this must give the bat valuable information about the movements of the insect it is pursuing.

SIZE OF BATS' EARS
As might be expected, bats' ears are extremely sensitive and very large. In the long-eared bat they are longer than the combined length of the head and body. They therefore form

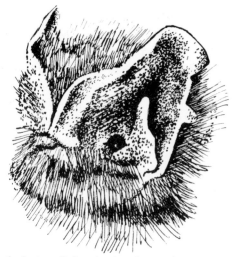

Head of a barbastelle bat showing the prominent tragus

very efficient ear trumpets for collecting the reflected sound. Our ears have a small fleshy lobe in front of the ear hole known as the tragus. In most bats the tragus is much enlarged and stands up like an additional ear flap in front of the ear hole. It is believed to play an important part in focusing the ultrasonic waves into the ear hole. Only in the horseshoe bats is the tragus either absent or reduced to a mere remnant.

FISH-EATING BATS

In America there is a small group of fish-eating bats that live on small fish which they snatch out of the water with the powerful claws of their hind feet. They fly at dusk, and like most other bats produce ultrasonic squeaks, but it seemed unlikely that these could penetrate the water and return as effective echoes. Calculations showed that owing to absorption by the water only 0.1 per cent of any sound entering the water would return through the surface as an echo, which would thus be extremely difficult to detect. Observation of these bats at work showed that what they in fact do is to detect any fish which breaks the surface, and swoop down on it. They tend to work with pelicans when these are fishing, catching the fish which break through the water surface when they are fleeing from the birds. They also take advantage of the fact that when shoals of small fish are being pursued by larger predatory species, some of them will break the surface in their efforts to escape.

VAMPIRE BATS

Apart from these fishing bats all the bats we have so far considered are insectivorous, using their ultrasonic squeaks to catch flying insects at night, lying up in caves, belfries and other suitable hideouts during the day. There is one other important group of nocturnal bats, the vampires. These have abandoned insect food and instead feed on the blood of other animals. They are very skilled at settling on sleeping prey, including man, and nicking the skin with their razor-sharp teeth, lapping up with their tongues the blood that flows freely from the wound. They do this so gently that their victims seldom wake up. Vampire bats also find their way about using ultrasonic squeaks of very high frequency, 100,000cps or even more. These squeaks, however, are so faint that they can only just be detected by instruments which easily detect the squeaks of other bats. For this reason they are sometimes referred to as whispering bats. It is a known fact that dogs are very seldom bitten even in areas where vampire bats are very common. The probable reason is that the sensitive ears of dogs are able to hear ultrasonic vibrations, so an approaching vampire bat will wake them up.

THE TOMB BAT

The largest of all bats are the tropical fruit bats which fly about during daytime in search of the fruit on which they feed. They use their well-developed eyes to find their way about and an acute sense of smell to locate the fruit. They do not therefore need to produce ultrasonic squeaks. If they are blindfolded they are unable to fly. There is, however, one exception, the species known as the tomb bat (*Rousettus aegyptiacus*), which goes to roost in tombs, temples and caves where it is completely dark. When these are flying about in daylight they use their eyes for navigation, but as they enter their dark hideout at the end of the day they are able to switch over to ultrasonic navigation, producing squeaks with frequencies as high as 100,000cps. These squeaks, however, are not produced by the voice as they are in other bats, but by slapping the tongue sharply on the roof of the mouth and making a kind of 'tut-tut' noise, each 'tut' consisting of a series of ultrasonic vibrations.

MOTHS AND BATS

Between bats and at least certain of the night-flying moths on which they prey, there has evolved a battle of ultrasonic wits. These moths have a pair of primitive ears, one on either side of the thorax just before it joins the abdomen, and these ears are capable of picking up the bats' ultrasonic squeaks. If the bat is some way off the moth will take evasive action, flying first this way and then that, in hope of eluding its pursuer, just as a bomber in wartime caught in enemy searchlights will take similar evasive action. If these initial tactics are unsuccessful and the bat begins to close in, the moth plays its trump card, folding its wings and plummeting down to the ground.

A few species of moth have developed an even more sophisticated method of confusing their enemies. As soon as they detect an approaching bat they are able themselves to produce ultrasonic pulses which effectively jam the bat's own navigational devices, and make it impossible for it to locate and catch the moth.

Even this, however, does not represent the ultimate protective achievement of night-flying moths in their eternal battle with their bat predators. In 1968 an American biologist, Dorothy Dunning, discovered that certain moths are repulsive to bats, who spit them out immediately if they happen to catch them. As a warning to the bats the moths emit their own ultrasonic signal, so that once a young bat tries to eat one it learns to avoid these particular species — which in future it will recognise from the ultrasonics — for the rest of its life. So for the sacrifice of a small number of individuals these species of moths gain immunity from attacks by bats. Another group of moths has learned to profit by this protection. Although they themselves are quite palatable, they have learned to produce ultrasonic signals similar to those produced by the unpalatable moths, thus leading the bats to identify them with the distasteful species, and therefore to avoid them.

OIL BIRDS

The work of Griffin and Galambos on echo location in bats stimulated investigations which revealed that other animals, too, use reflected sound for navigational purposes. Griffin himself went to Venezuela to study the curious oil birds — nocturnal birds about the size of a crow — which roost during the daytime in deep caves where it is completely dark, coming out at dusk to feed through the night. Their food consists of oily fruits and nuts, and their bodies are so full of oil that in earlier times they were collected in large numbers from the caves and boiled down to produce oil for use in lamps. Oil birds had long been known for the continuous twittering which went on all the time they were flying. These sounds are produced in the range 6,000 to 10,000cps, and are thus well within the range of the human ear. Griffin was able to show that the birds used the echoes received from these audible sounds to find their way about in the same way as bats use their inaudible ultrasonic vibrations. In the caves he was able to pick out both the sounds and their echoes. So accurate is the bird's navigation that it is able to locate its own nest among hundreds and maybe thousands of others surrounding it in complete darkness.

SWIFTLETS

Another bird which nests in pitch-dark caves and finds its way about inside them is the little swiftlet of Borneo, Malaya and other parts of the Far East. Unlike the oil bird, the swiftlet leaves its cave to feed during the daytime, making use of its good eyesight to navigate. At these times it is generally silent, but at the end of the day when it returns to roost it switches on a continuous series of squeaks similar to those produced by the oil birds the moment it flies through the cave entrance. These are well within the range of the human ear, and to anyone standing near the entrance to one of these caves as thousands of birds are returning the racket is quite astonishing.

If a number of these swiftlets are released into a completely dark room they fly around quite happily, never colliding either with the walls or with each other, twittering loudly all the time. If a light is then switched on the twittering ceases immediately as the birds switch over from audial to visual navigation.

Until their prowess at echo location was discovered, the swiftlets' chief claim to fame was the fact that they construct their tiny nests with their own saliva, sticking them to the bare vertical walls of the caves. These are the birds' nests which are considered such a delicacy by Chinese and other peoples living in the Far East, for they are the principal ingredient of the celebrated birds' nest soup. The natives who undertake the collecting of the nests risk life and limb swaying at the top of long flexible ladders made of bamboo. Cave space in the area is very valuable both to swiftlets and to bats, and most of the caves are occupied by enormous colonies of both animals. There is no overcrowding, however, for as the swiftlets return for the night, so the bats leave to spend the night outside the cave hunting for food.

The ability of oil birds and swiftlets to locate their own nest among thousands of others implies much more skill than merely avoiding collisions with solid objects. Each bird must carry with it an accurate memory of the exact echo pattern which it receives when its twitterings are reflected back to it from its own nest and the area immediately surrounding it.

MANX SHEARWATERS AND STORM-PETRELS

R. M. Lockley, the British naturalist, has spent a great deal of time studying the sea birds which nest on the island of Skokholm, off the west coast of Wales, and is convinced that the Manx shearwaters and storm-petrels which nest in great numbers on the island do, under certain circumstances, use echo location to find their nests when they come in from the sea.

They build their nests in rabbit holes in the ground, and because of the many predatory birds such as gulls, hawks, crows and others which would pounce on them if they returned to or emerged from their burrows during daylight, they will only come out at night. On moonlight nights they can also sometimes be detected and caught by their enemies, and so there is less returning and departing when the moon is full.

Those birds which return on bright nights do so silently, dropping down to their burrow entrances and disappearing inside as quickly as possible, as though aware of the dangers of lingering outside. On dark nights, however, returning birds approach the nesting sites screaming. It had always been assumed that under these circumstances the screaming of one bird served to alert its mate, which had remained on the nest with the eggs or the chicks, and that it would answer the call of the returning bird and help to guide it back to the right burrow. Lockley, however, was able to establish that often there was no reply from the mate on the nest. In fact after the eggs have hatched both birds are often away from the nest feeding and collecting food for the nestlings at the same time, so that there would be no one on the nest to give an answering call. Yet without any such guiding signal the birds were able to locate their burrow entrances with the same kind of uncanny accuracy with which oil birds and swiftlets are able to locate theirs in pitch-dark caves. Although not yet proved, it seems virtually certain that these shearwaters and petrels could only home on their burrows with such accuracy with the aid of echo location, of which the screaming forms a part.

ECHO LOCATION IN DOLPHINS AND OTHER WHALES

So far we have considered only animals which produce ultrasonic vibrations in air, but further research since the last war has established that echo location is used by a number of marine creatures. Water is in fact a better medium than air for echo location, because sound of any frequency travels much better through water than it does through air.

One of the major developments since World War II in the keeping of wild animals in captivity has been the establishment of oceanaria—giant tanks filled with sea water in which dolphins, sharks and other large marine animals can be exhibited. The first oceanarium was established in Florida, and its curator, the late Arthur McBride, made the first significant observations which led him to suggest that dolphins must employ ultrasonic frequencies to find their way about. He found that if he suspended a fine-mesh net across part of the tank the dolphins never swam into it, even on pitch-black moonless nights, or when the water was so turbid that visibility within it was less than one yard.

Further experiments were conducted by two other American investigators, W. E. Shevill and B. Lawrence. They placed discs over the eyes of dolphins and transferred them to a new tank in which a number of vertical wires had been strung. The dolphins were able to swim at speed around the tank without once touching the wires or colliding with the sides. They then placed a single blind-folded dolphin in a tank in which there was a single small fish of similar size to a herring. The dolphin was at once able to detect the swimming fish, chase after it and eat it. Other fish were then thrown in and caught as quickly as they would have been if the dolphin had not been blindfolded. Whether the water was clear or turbid made no difference to the dolphins' uncanny abilities. One dolphin was even able to distinguish between a piece of fish and a gelatine capsule filled with water of the same shape and size. It ate the piece of fish, but veered away from the capsule.

Systematic research after these preliminary demonstrations has pieced together the complete picture of dolphin echo location. Like bats, they produce intermittent pulses of ultrasonic vibrations, the frequency of which can be as high as 200,000cps, the pulse rate itself varying from 10 to 400 per second. This rate, as with bats, increases as an object is approached, to be immediately reduced once the object has been passed or eaten. When the dolphin is swimming and emitting these ultrasonic pulses it turns its head from side to side in order to scan as wide a field ahead as possible.

All whales have their nostrils on top of the head, and between these and the tip of the snout is an area known as the melon. Now although dolphins offer no resistance to their eyes being covered with rubber caps, they will not tolerate this melon being covered. And if a microphone is suspended in the water when a dolphin is swimming, the vibrations which are picked up are strongest when the melon is pointing towards the microphone. These facts have led to the suggestion that the melon probably acts as a focusing device for the emitted vibrations, as does the nose-leaf of the horseshoe bats.

All the observations so far made have been conducted with the toothed whales, because it is this group which contains the relatively small dolphins and porpoises which can be kept easily in captivity. It is likely, however, that the large whalebone whales also navigate by producing ultrasonic frequencies. During the last war it was observed that when warships were using ultrasonic vibrations (Asdic) for submarine detection, any whales in the area, toothed or whalebone, were scared away.

Incidentally all whales produce a great variety of audible sounds, from high-pitched whistles and trills to noises resembling creaking doors. These sounds, however, have nothing to do with navigation. They are conversational noises, and help the members of a school of whales to maintain contact with one another as they roam the oceans.

ECHO LOCATION IN SEALS AND PENGUINS

Subsequent to the discovery that whales can navigate by ultrasonic vibrations it was shown that at least some of the seals and sea-lions can also navigate by echo location. In a series of experiments fish were thrown into a tank containing sea-lions, the water having been made so turbid that the sea-lions must have been working in complete darkness. Yet despite their inability to see they caught fish as quickly as they would have done in completely clear water.

Penguins too, it seems, can catch fish in complete darkness, but whether they use echo location has yet to be established. It has been suggested that they pick up the echoes from the strong turbulence they cause as they swim through the water. Only future research will finally show how they catch prey which they cannot see.

CHAPTER 4

MIGRATION AND ORIENTATION IN LAND VERTEBRATES

True migration is a comparatively rare phenomenon among land vertebrates. This is not really surprising, because it is much more difficult for a land animal to cover great distances on foot, with the likelihood of having to cross rivers, mountains and deserts, than it is for aerial and aquatic animals to achieve similar distances through air or water. In any case it would only be the larger kinds that could travel far enough for the migration to be of any use to them.

The main possible reasons for migration are to avoid severe winter climates, to reach better feeding grounds at certain times of the year, or to visit suitable breeding grounds at the breeding season. In no part of Africa is there a severe winter climate from which an animal might want to escape, and in Asia a virtually insurmountable mountain barrier separates the warmer south from the cold north of the continent. Only in North America, with its arctic northern regions not separated by severe barriers from the warmer southern areas do we find conditions in which we might expect to find some migration of larger animals.

NORTH AMERICAN CARIBOU AND BISON

The best known examples of migration among the North American mammals is the caribou, the North American species of reindeer. But even here there is no constancy about their migrations. In some years great herds assemble together for a mass migration, whereas in other years they migrate in such small numbers that the migration can be almost undetected. In the days of its abundance, the North American bison too moved south in their millions in the autumn to return in the spring. But since the herds were virtually wiped out by the 1880s there is now no annual migration of bison.

BREEDING MIGRATIONS OF AMPHIBIANS

Among the vertebrates, the group showing the most widespread ability to migrate and navigate accurately are the amphibians — the frogs, toads, newts and salamanders. They are called amphibians because although they can spend most of the year on land they must return to the water to breed each spring. During the winter frogs and toads living in temperate climates go into hibernation and during this time their reproductive organs ripen. When they wake up early the following spring they set off at once to find a pond in which to breed. This is not any old pond, but their home pond, where they themselves were originally hatched, and where they return year after year to breed.

They may well have hibernated the previous autumn a long way from their original home, and on their way back they may pass other ponds, even swimming through them rather than hopping round them. How do they recognise their own pond, and how are they able to distinguish it from all others? It has been suggested that each pond has its own particular smell, due perhaps to the particular assortment of plants and animals living in it. This may well be the correct explanation, for we have already seen in Chapter 1 that salmon can certainly recognise their own river by its smell.

Salamanders, too, return to their home pond each spring. Professor Twitty, an American zoologist, marked 260 salamanders which had returned to a pond to breed one spring. Each subsequent spring he kept a watch on the pond and found that a large number of them returned year after year. Even after seven years nearly one-third of those originally marked reappeared. This is an amazingly high proportion, because many of the original band must have fallen prey to other animals or have died of natural causes during the intervening years.

TRACKING ABILITIES OF THE DOG

Whether or not scent plays an important role in the spring migration of amphibians, it certainly plays a major part in the lives of many mammals. The species we know most about in this connection is the domestic dog. Because our sense of smell is so rudimentary compared with that of the dog we tend to marvel at the dog's astounding ability. But the dog, were it capable, would marvel equally at what it would consider our astounding sense of sight. The dog lives in a world of scent and sound, whereas we live in a world of sight and sound. Nevertheless the dog's sense of smell is remarkable by any standards.

As Dr Maurice Burton, British writer and naturalist, has pointed out, the olfactory or smelling membrane in our nose covers an area about the size of a postage stamp and contains about 5 million sensory cells which are sensitive to molecules which settle on them. The olfactory membrane of a dog of medium size if spread out flat would occupy an area fifty times greater, and in those dogs which are used for tracking this membrane may contain as many as 220 million sensory cells.

The sense of smell differs in one very important respect from the senses of sight and hearing. The latter depend upon vibrations or wave transmissions — light radiation and sound waves. Nothing physical travels from the source of the light or sound and the eyes or ears which detect them. Smelling, however, occurs only when actual molecules of a substance travel through the air from the substance and make physical contact with the olfactory membrane. Only in vertebrates, however, are the olfactory membranes contained in an organ called a nose. In other animals, as we shall see, the receptor cells are situated on other parts of the body.

To understand how a tracker dog can follow the scent of a human being we have to realise thàt, even when we are wearing shoes, at each step we deposit millions of sweat molecules on the ground. Some of these are molecules of butyric acid, to which the dog's olfactory cells are extremely sensitive. A bare-footed animal will likewise deposit tell-tale molecules. One of the earliest experiments designed to show the acuteness of the dog's sense of smell was conducted by G. J. Romanes in 1885. Twelve men, including Romanes himself, set off on a walk, Romanes leading. Each man carefully placed his feet in the footsteps of the man in front, so the scent particles deposited by Romanes's shoes were overlaid by eleven other sets of particles. After a time the group divided, Romanes and five of the men continuing in one direction and the remainder in another until each group reached a hiding place where they could not be seen from the routes they had taken. They had continued, of course, to tread in one another's footsteps. Romanes's dog was then released. It soon picked up the starting point, and could clearly recognise its master's scent from among the eleven other scents. Where the two tracks parted there was but a moment's hesitation before it set off along the correct track which brought it straight to its master.

The eventual aim of Romanes's experiments was to find out whether a dog was able to distinguish between identical twins by scent. He repeated the same experiment using a pair of identical twins among the twelve people involved. At the dividing point one twin went with five of the others, and the other twin with the remainder. Before the experiment started the dog was presented with the scent of one of the twins. In the event, although the dog was able to follow the joint trail left by both of them, it was unable to distinguish between the two twins when the trail divided. It was just as likely to follow the trail left by the wrong twin as the right one, though it would never confuse one of the twins with anyone else. Subsequent research showed that identical twins have very similar if not identical scents.

Dr Burton records an occasion when the Cairo police wished to trace the owner of a donkey. He was known to have ridden the animal along a rocky track as much as four days before, nevertheless one of their trained Alsatians was able to pick up the scent and lead them without hesitation to the house in which the donkey was found.

THE RATTLESNAKE'S INFRA-RED 'EYES'

Many animals, as we shall see in later chapters, use radiant heat (infra-red radiation) emitted by living bodies to locate their victims. One group of snakes are exceptionally well adapted to do this. They are the group known as the pit-vipers, because on each side of the head between the eye and the nostril there is a tiny pit a few millimetres across and a few millimetres deep. The best known of these pit-vipers are the rattlesnakes, but the group contains also the moccasins and certain other species. The possible function of these pits was the subject of conjecture for a long time, but it was though that they must play some sensory role.

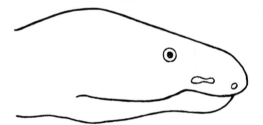

Head of rattlesnake showing the position of the pit between the nostril and the eye

In the end, examination of the pits showed that at the bottom of each was stretched a membrane consisting of enormous numbers of cells very similar in structure to the heat receptors found in the human skin—the receptors which enable us to detect the warmth of an object when it comes into contact with our skin. But whereas we possess an average of 3 receptors to each square centimetre of skin surface, their concentration on the pit membrane is at the rate of 150,000 per square centimetre, or something like 15,000 in each tiny pit organ.

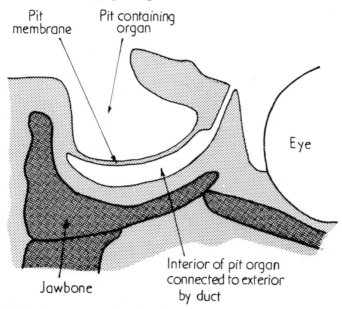

Section through a rattlesnake's pit organ showing the pit membrane containing about 15,000 heat-sensitive cells which enable the snake to detect objects whose temperature is only a fraction of a degree higher than that of surrounding objects

Such a vast concentration makes the pit organs remarkably sensitive to radiant heat. They are capable of locating an object whose temperature is just a fraction of a degree above that of the general surroundings. And since all living things, warm-blooded or cold, give off some radiant heat, the pit organs provide the snakes with an extremely sensitive method of detecting their prey at night.

Experiments carried out in 1952 by Professor T. H. Bullock, an American physiologist, proved conclusively the value of its pit organs to the pit-viper. In his first experiment he blindfolded a rattlesnake with adhesive tape and sprayed the inside of its mouth with a substance which would deaden its ability to taste or smell. He then put it into a vivarium with a live mouse. It behaved as though it could see the mouse clearly, following its every move with great accuracy. Within a short time the mouse had been captured and was being eaten.

He now took the same rattlesnake and covered its pits as well with adhesive tape, and then released it again in the vivarium. This time he put in several mice with it. After several days not a single mouse had been caught or harmed. Clearly with its pits covered the snake had been 'blinded'. Normally its pits function as infra-red 'eyes'.

At night, when rattlesnakes do much of their hunting, they glide along moving their heads from side to side, scanning the environment ahead for infra-red rays. When an animal is detected the snake is able to estimate its size by the side to side movements of its head, and so decide if it is too big to tackle. Even during the daytime, when the snake is hunting through undergrowth, the pits still come in useful. Many of the kinds of animals on which it preys are camouflaged, and might well escape detection by the snake's not particularly efficient eyes. But they give themselves away by the radiant heat that their bodies are emitting.

CHAPTER 5

THE FACTS OF INSECT MIGRATION

The facts of bird migration have been known for a very long time, but until fairly recently it was thought that insects did not migrate. True migration, as we have seen, involves travel from one area to another at a certain time of the year, and a return journey at another time. That certain butterflies, moths and other insects sometimes undertook journeys over considerable distances in enormous numbers was known for a long time, but these were not accepted as migrations, because there was no evidence that any members of these swarms subsequently made the return journey. This lack of evidence, we now know, was due to lack of observation, and over the past few decades a great mass of information has been accumulated about insect migration.

PAINTED LADY AND MONARCH BUTTERFLIES

Migration among insects occurs most frequently in butterflies, some of which are able to travel immense distances. One of those whose migratory abilities we know most about is the painted lady (*Vanessa cardui*), whose distribution is almost world-wide, South America being the only continent where it is not found. All the painted ladies of Europe and North America are believed to have been derived originally from one single vast population. During the late autumn and winter they breed along the northern fringes of the Sahara desert. By March they are emerging from their chrysalides by the million,-and after pausing for an hour or

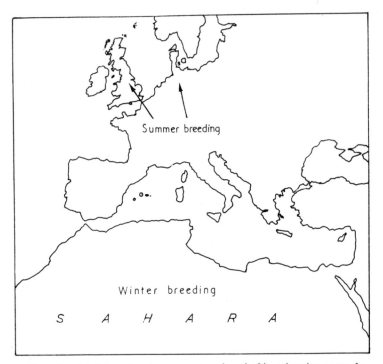

The painted lady breeds twice a year at each end of its migration range. In winter it breeds at the southern limits of its range on the edge of the Sahara Desert, and in summer it breeds in Europe

two for their wings to dry they fly off northward. The north African coast when they reach it is no deterrent, and they fly out over the Mediterranean. This might seem a very wide stretch of water for butterflies to be able to cross, and indeed for a long time it was thought that even the English Channel would be too great a distance for such insects. But we now know that butterflies are able to fly much greater distances across the sea. Once they reach southern Europe the painted ladies continue to fly northward, but from central Europe northward numbers of them settle and breed. In most years the limits of the migration are southern Scotland and northern Germany, but in exceptional years considerable numbers may fly as far as northern Finland and Iceland.

By the autumn the eggs laid by these migrants will have produced the next generation of adult butterflies, and these now set off on a southerly return migration which continues until they have crossed the Mediterranean and have reached the borders of the Sahara. Here they mate, and it is their eggs which eventually give rise to the adult butterflies which emerge from the chrysalides the following March. The distance covered by butterflies which emerge at the edge of the Sahara and eventually reach Britain is at least 2,000 miles.

There is one important difference between the migration of painted ladies and birds such as swallows and martins. The birds only breed once a year at one end of their migration range, whereas the butterflies breed twice a year, at both ends of their range.

Another significant difference between bird and butterfly migration in general is that whereas the number of birds of any species migrating year by year remains fairly constant unless there is some climatic or other catastrophe, there is very considerable variation from one year to another in the number of migrating butterflies of any species. One of the most impressive flights of painted ladies was observed in California in 1924. It was estimated to consist of something like 3,000 million butterflies, all flying in the same direction.

Before it was established that butterflies really did migrate it was assumed that their mass movements were merely the result of them being carried by the prevailing winds. We now know, however, that butterflies on migration are able to keep to a definite course even if this carries them against or across whatever wind may be blowing. It seems, too, that each butterfly is able to maintain its course even though it cannot see any of its migrating fellows. Steady migration of painted ladies across the

Mediterranean has been observed at the rate of only twelve butterflies an hour. Since they fly at about six miles an hour, this means that the gap between two consecutive butterflies must have been about half a mile, and they could not possibly keep one another in sight over this distance. Yet they all maintained the same course. On other occasions individual butterflies flying northward have been observed to approach a ship sailing through the Mediterranean, fly up over one side, across the deck, and down the other side to continue their journey without changing course.

The monarch or milkweed butterfly

Another butterfly whose migrations are well-known is the monarch or milkweed butterfly (*Danaus plexippus*), which is common and widespread in North and South America. It owes its alternative name to the fact that its caterpillars feed only on the plant known as milkweed. More is known about the North American than the South American populations. Through the summer months monarchs are common everywhere in the USA and southern Canada, and in some years they can be found as far north as Hudson Bay, and everywhere they are breeding. Towards the end of August in Canada, and somewhat later in areas further south, the new-generation butterflies begin to leave the areas where

they were bred during the summer and fly steadily southward. The parents have by this time completed their life's
work and have died.

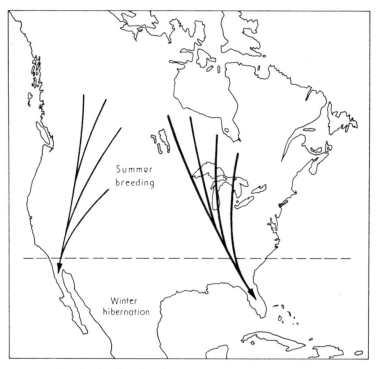

Map showing the migration range of the monarch butterfly

At first they fly in small groups each containing only a few
butterflies, but as they fly on these smaller groups tend to
come together to form larger groups until eventually large
swarms may be formed. One such swarm was observed one
evening to have occupied every twig on every bush and tree
over an area 200yds wide and 2 miles long. The total
number of butterflies must have numbered many millions.
Early the following morning they had all disappeared,
continuing their southward migration.

These mass movements of monarch butterflies do not take place across the whole country. The majority fly down the eastern half of the USA and there are considerable migrations down the west side near the Pacific coast; but down the mountainous middle west there are no migration flights. The eastern flights reach Florida and the western flights reach California before finally coming to rest sometime during October. Here they settle on trees and go into hibernation until the following spring, although during warm sunny days they may wake up for a short time and flutter around. There may be thousands of butterflies on a single tree, and the same groups of trees may be used year after year. Some sixty miles south of San Francisco is a wooded area called Pacific Grove, which is noted for the immense numbers of butterflies which are known to have come there to hibernate every year for at least the last eighty years. So spectacular is the sight of the myriads of butterflies on the 'butterfly trees' that they have become a tourist attraction. There are severe penalties for anyone throwing stones into the trees or in any way disturbing the hibernating butterflies.

As the days begin to get warmer towards the end of March the butterflies begin to wake up and fly off northward. But whereas in the autumn they collected together and migrated in conspicuous mass flights, they now fly alone. This of course makes the spring migration much harder to detect, and explains why it was not discovered for a long time. At this time, too, the new shoots of the milkweed are beginning to show above ground in the southern States. Some of the monarchs make only a short migration before mating, laying their eggs on the milkweed, and then dying. This first generation can usually be succeeded by a second and a third generation before they need to set off on the short southward migration towards the end of September or early in October. Those butterflies, however, which continue their spring flight to the central or northern States will produce their first generation offspring about a month later than those which remain in the southern States, leaving time for only a second generation to be produced before the time comes for them to migrate southward early in September. Those which continue into Canada have only time to produce a single generation, because the spring and autumn migrations both occupy about two months.

There are two important differences between the migrations of painted ladies and monarchs. Painted ladies breed at both ends of their migration range, monarchs only during the summer at the cooler end of their range. Monarchs, too, are unusual in that their life cycle includes a period of hibernation. They are not, however, the only insects to hibernate, as we shall see.

BUGONG MOTHS

Much less is known about migration in moths because the majority of them are nocturnal, and it is much more difficult to detect migration flights at night. Nevertheless we do know that some moths at least undertake regular mass flights from one area to another, among them many of the hawk moths. One of the few species whose migrations are known in detail is the Australian bugong or bogong moth (*Agrostis infusa*), which is found mainly in Victoria and New South Wales. In early spring, around the beginning of September, the moths are quite common on the plains, but towards the end of October they begin to disappear, and for the next few months, through the height of the summer, none can be found. From early March, however, they begin to reappear, and within a few weeks are once again abundant. It is now, at the beginning of the southern autumn, that they breed and give rise to the next generation of moths which will disappear in their turn the following spring.

Investigations which were begun in 1950 have now pieced together the whole story. When the moths leave the plains in the spring they fly into the mountains, reaching altitudes of up to 6,000ft or more, and here they undergo a kind of summer hibernation or aestivation, hiding away in caves and rock crevices. True hibernation is a state of suspended animation in which some animals, like monarch butterflies, pass the colder months of the year, whereas aestivation is a similar kind of suspended animation enabling an animal to survive the hottest months of the year. Immense aggregations of the moths were first discovered by an explorer on mountain tops in New South Wales as long ago as 1833, but the native Australians must have known about them long before this, because it was even then traditional to make treks into the mountains to collect them for food during the summer months. The concentrations of these aestivating moths can be quite remarkable. As many as 500 individuals can be found on a single square foot of rock surface in the crevices and caves.

There is a remarkable relationship between the bugong moth and two species of parasitic worms. These are found only in the moths' mountain retreats. As the moths arrive in the spring to aestivate the young worms have just hatched, and they invade the tissues of the moths, in whose bodies they then proceed to develop during the ensuing months. Before the moths are due to resume an active life and return to the plains the worms, now well developed, leave their bodies and burrow into the soil on the floor of the caves and crevices. During the following winter they complete their development, mate, and lay their eggs, from which the young worms emerge in time to infect the following year's moths as it comes to their turn to retreat to the mountains to aestivate. Neither of these worms exists on the plains, so if the moths changed their habits and ceased an annual migration to the mountains the worms would become extinct.

ARMY CUT-WORM MOTHS

Another moth which makes aestivation migrations to the

mountains for the summer months is a North American species known as the army cut-worm (*Chorizagrostis auxiliaris*). Like the bugong moth, the cut-worm is common in the spring and then disappears completely during the summer months, to reappear in the autumn. The spring moths are sexually immature, but those appearing in the autumn are ready to breed. Only recently has it been discovered that, as summer approaches, the moths migrate to the mountains to aestivate at heights of up to 9,000ft. Apparently grizzly bears know about these accumulations of moths, because scientists investigating the habits of the bears have found that in the mountains their summer droppings are very rich in the remains of the cut-worm moths.

THE 'SUNN PEST'

One insect, a plant bug known as the 'Sunn pest' (*Eurygaster integriceps*), has a life-history involving both aestivation and hibernation. It lives in the Near East, where it is a serious pest of spring barley crops. As the barley begins to sprout in the valleys in the spring so the bugs suddenly appear in swarms, not to feed on the barley but to mate and lay their eggs, after which they die. Within a few days myriads of young bugs hatch out and proceed for the next month or so to gorge themselves on the tender shoots of the young barley, causing severe devastation unless they are controlled by spraying. At the end of this time they have not only become full grown, but have laid down in their bodies sufficient stores of food to last them for the remaining eleven months of their lives, for they will never feed again. By now the temperature in this part of the world is increasing rapidly, so the insects fly up from the valleys into the mountains where it is cooler, and, at a height of 7,000-8,000ft, they enter a period of prolonged aestivation. They become active again in the autumn, when the temperatures are beginning to fall, and now undertake a second migration which brings them down to about 5,000ft. They now go into a hibernation which will carry them right through the winter. In the spring they once more resume activity and migrate for the third time, arriving in the valleys, as we have already seen, just as the spring barley is beginning to sprout.

LOCUSTS

The classic example of insect migration, which has been known for thousands of years, is of course that of the locust. The first account is given in the book of Exodus, believed to have been written about 1500 BC. In its essentials it is an accurate account of locust migration and devastation as we know it today. The locusts came in with the wind and 'covered the face of the whole earth', with the result that 'there remained not any green thing, either tree or herb of the field, through all the land of Egypt'. The fact that a swarm soon moves on after it has done its worst is also recorded. 'The Lord turned an exceedingly strong west wind which took up the locusts and drove them into the Red Sea.'

Locusts are just rather large kinds of short-horned grass-hoppers, some of which undergo spectacular but irregular migrations. Of the seven kinds known, two are found in America, the other five being Old World species. From the point of view of migration, though, there are two species which are more important than the others, the desert locust (*Schistocerca gregaria*), which is found mainly in and around the north African desert belt and in the Near East, and the migratory locust (*Locusta migratoria*). This is much more widespread than any other species, occurring almost everywhere in Africa, in southern Europe, and extending through southern Asia to Australia and New Zealand.

A typical swarm of desert locusts observed by Dr C. B. Williams on 29 January 1929 at Amani in north-eastern Tanganyika was more than 1 mile wide, at least 100ft deep and took 9 hours to pass, the locusts flying at about 6 miles an hour. At an absolute minimum Dr Williams estimated the number of insects in this swarm as 10,000 million. They arrived during the morning and settled, breaking branches off trees up to 3in diameter with their combined weight. During the following morning they left, and there was not a single grass blade or leaf to be seen: the whole area had been completely denuded of foliage. I quote this particular example because it was observed and recorded by one of the world's leading authorities on insect migration, and is therefore likely to have been as accurate as scientific investigation could make it.

Most insects on migration flights fly at the same height above the ground as they do in normal daily flights, which means usually less than 10ft, although monarch butterflies on migration have been seen flying much higher than normal at up to 400ft above the ground. Migrating locusts, however, usually fly very high, and swarms have been encountered by aircraft as high as 7,000ft. This has an important bearing on the direction of migration. Most insects can determine the direction of their migration, being able to fly against or across the wind as well as with it. They are able to do this because at ground level wind speed is not very great. But high in the air the wind speed is much greater — certainly greater than a locust can fly — so that once a swarm has risen far above the ground it will be carried in the direction of the prevailing wind. A locust swarm is therefore incapable of determining its own direction of migration.

Most of the so-called migrations of locusts do not in fact fit in with the definition of migration adopted in this book. As explained earlier in the chapter, for a journey undertaken by a large number of animals of a particular species to be regarded as a migration there must be evidence of a return journey, either by those individuals which made the original journey or by their offspring. Many of the locust swarms certainly do not undertake a return journey.

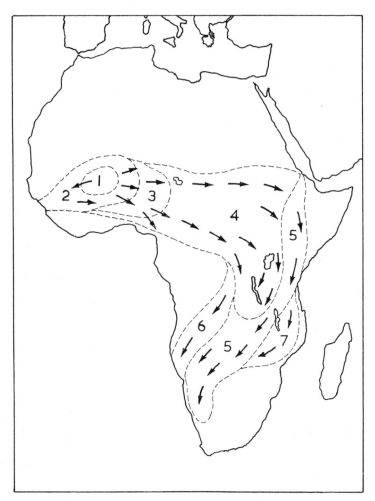

Map showing the gradual increase and spread of the migratory locust in central and southern Africa between 1928 (1) and 1934 (7). The intervening numbers indicate where the locusts reached in the intervening years

Nevertheless these movements of gigantic numbers of locusts must be included in this survey because they do constitute a special kind of migration. There is in fact some evidence that in north Africa the desert locust does indulge in regular

86

seasonal migrations. In the winter there is a brood which is produced to the north of the Sahara. By the spring these have become adult, and fly southward across the desert. South of the desert they breed, and their offspring in their turn fly northward in the autumn ready to produce the winter brood. There are thus two generations a year.

The general pattern, however, in any particular area is for the locusts to be unnoticed for a number of years, and then for there to be a sudden enormous swarm, often followed by others in succeeding years, after which once again they seem to fade into the background. For a long time the cause of these periodic locust outbursts followed by their apparent disappearance was unknown. For an explanation of the mystery we are indebted to Dr B. P. Uvarov, who began an extended investigation of the locust in 1921 as head of the Anti-Locust Research Centre based in London.

He began by comparing the migratory locust (*Locusta migratoria*) with a solitary grasshopper of similar size (*Locusta danica*). There were significant structural differences between the typical members of the two species which made it easy to distinguish between them, and there was also a colour difference. The locusts were dark, whereas the grasshoppers were pale-green. Besides the typical members, however, Dr Uvarov was puzzled by the frequent occurrence of intermediate forms which could not be placed with certainty in either species. Another fact which interested him was that the geographical distribution of the two species seemed to be virtually identical. Eventually he was able to show that the locust and the grasshopper were not in fact two separate species, but two different forms of the same species, one form being solitary and non-migratory, the other gregarious and liable to produce immense migratory swarms. The different forms Dr Uvarov called phases, so the migratory form became *Locusta migratoria* phase *gregaria,* and the solitary grasshopper form *Locusta migratoria* phase *solitaria.* The intermediate forms which had so puzzled him at the start of his investigations became the *transiens* phase.

The final demonstration of the truth of his theory was crucial and simple. He found that if he took a batch of eggs laid by a single female migratory locust, separated them into two smaller groups, and then allowed one group to develop under crowded conditions, it gave rise to typical migratory locusts. The members of the second group were isolated from each other throughout their development, and they gave rise to typical pale-green grasshoppers. Subsequently it was shown that other species of migratory locusts could also exist in the solitary grasshopper phase. It was also discovered that locusts could migrate even in the *solitaria* phase, sometimes covering distances of several hundred miles. But such migrations may consist of much smaller numbers than the true migratory swarms, and they are not so devastating.

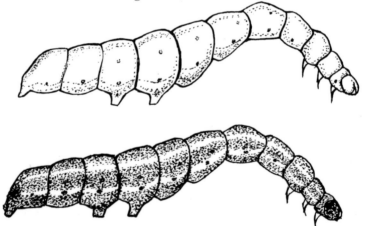

Colour phases in the larvae of the silver-Y moth. Normally the larvae are light in colour, but during severe outbreaks when overcrowding results they take on a dark colour

Since the discovery of phases in locusts it has been found that certain other insects can exist in two different colour phases which are related to the incidence of crowding or relative isolation during development. In 1943 it was demonstrated experimentally that the larvae of two south African moths, the army worm and the lesser army worm, developed a much darker colour and were much more active if they were kept crowded together as soon as they hatched. Ten years later similar results were achieved in crowding experiments with the larvae of the common European silver-Y moth (*Plusia gamma*). During severe outbreaks of this moth in the wild similar dark-phase larvae have been found.

Dr Uvarov's discoveries forced a change in ideas about the means by which migratory locusts might eventually be exterminated. Prior to this it was believed that if every swarm, wherever it occurred, could be destroyed, the locust population would gradually be reduced until total destruction could be achieved. But he showed that whenever a swarm is almost destroyed, the eggs laid by the few stragglers which are bound to survive develop into the *solitaria* phase. These insects are inconspicuous and can go on multiplying in the area for some years without being noticed. Eventually, however, there will come a year when climatic conditions are particularly favourable to their development, and enormous numbers of eggs are laid. These hatch under crowded conditions and therefore develop as *gregaria*-phase locusts, which will fly off as a swarm when they have reached maturity. Thus the only hope of exterminating a locust species would be to kill off all the *solitaria*-phase individuals, a much more difficult task even than destroying the *gregaria* swarms, because they are so much more difficult to locate.

The life-history of the locust is bound up with that of the solitary wasp. The latter are particularly interesting to the naturalist because of their ingenious method of providing food for their larvae. Most of them dig holes or burrows in the ground and deposit a single egg in each one. They then capture small animals which they paralyse by stinging, and drop one into each burrow before sealing it up. Each species has its own particular kind of victim, which may be an adult insect, a caterpillar or a spider. A number of them choose various species of grasshoppers and locusts. *Sphex aegyptius* is an east African species which preys upon the desert locust. When a swarm of locusts flies off on its migration journey, the wasps migrate with it, and when it settles the female wasps set about excavating holes and paralysing locusts with which to fill them, continuing without rest so long as the swarm remains. But when the area has been devastated and the locusts take to the wing once more, the wasps cease whatever they are doing and fly off with them, leaving large numbers of unfinished burrows and many paralysed locusts lying around which would have been used used to fill them had there been sufficient time.

DRAGONFLIES
Besides butterflies, moths and locusts, a number of other different kinds of insects are known to undertake mass flights which often involve enormous numbers of individuals travelling considerable distances. For many of these there is as yet no clear evidence of subsequent return flights which would enable us to class them as true migrations. But, as already mentioned, compared with other animals which migrate, mass movements of insects occur only periodically with no apparent movement in the intervening years. It seems likely that further knowledge will show that most mass movements of insects are indeed examples of true migration involving a subsequent return. We have seen that the southward autumn migration of monarch butterflies is conspicuous because large numbers fly together, whereas the return spring flight is difficult to detect because then the butterflies migrate as solitary individuals, so that until recently the southward mass flights were not regarded as migrations.

Mass movements of dragonflies have been known for a very long time. The earliest recorded movement was a north to south mass flight reported in Germany in 1673, and since then a large number of such flights have been recorded. One of the most spectacular recent flights was recorded on 2 and 3 September 1947 crossing the south coast of Ireland flying in from the sea. At the time no flights were recorded into south-west England or into France, and it is believed that these dragonflies had flown across the Bay of Biscay from Spain and Portugal, a distance of about 500 miles. It is reported that the concentration of insects was so great that it frightened people. Similar mass flights of dragonflies have been reported from many parts of the world, some in a general northerly, others in a southerly, direction. In some places the sudden appearance of a mass flight of dragonflies heralds the subsequent arrival of a very high wind from the same direction, and is taken as a warning to make preparations to minimise the damage such a wind might do.

There is an interesting connection between dragonfly migrations and a disease of domestic chickens and turkeys. It is caused by trematode worms, whose effect is to prevent the birds from laying eggs, and eventually to kill them. As long ago as the 1880s regular mass migrations of dragonflies were known to occur periodically flying south-west along the Baltic coast of Poland. Whenever these swarms appeared the inhabitants hastily shut their hens away, since it was well known that after eating dragonflies they ceased to lay. Why this should be so, however, was a mystery. This trematode disease was widespread in Europe, North America and the Far East, and was known also to affect certain aquatic birds. Most trematodes have an invertebrate intermediate host in which the larval parasites develop and a systematic search was made for the intermediate host of this particular flatworm. The first step in solving the problem was made in 1929 when Dr Siskoff, a Russian scientist, was able to relate an outbreak of the disease to the sudden appearance of large numbers of dragonflies. Two years later a colleague of his, Dr Szidat, was able to identify large numbers of the intermediate stages of the trematode parasite in the dragonfly nymphs.

LADYBIRDS

Ladybirds are another group of insects which are well known for making mass flights over considerable distances, but here again sufficient evidence is lacking at present to

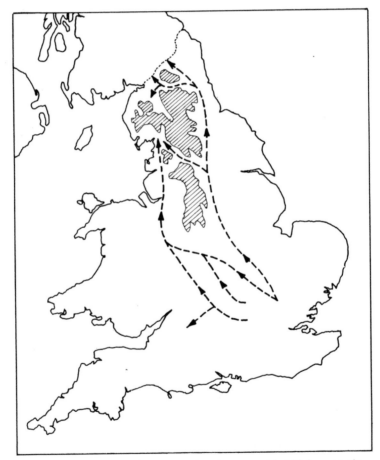

Map showing the migration routes taken by two-spot ladybirds across the Pennines in June 1925

show that all of them are true migrations. In some cases it is known that these flights represent journeys to and from places where the ladybirds hibernate. In Britain there are forty different species of ladybirds, and six of these appear in mass flights from time to time, the most important of these being the two-spot, seven-spot and eleven-spot species. There are many records of the sudden appearance of immense numbers of ladybirds flying in from the sea over the south and east coasts. Usually in such cases the appearance of the swarm is recorded only from one locality, and there are no records from other localities which would give a more complete picture of the migration.

There is, however, one example in which an extensive migration was recorded by more than fifty observers in various parts of its range. In June 1925 immense numbers of the two-spot ladybird (*Coccinella bipunctata*) appeared all over the north of England. Marriner collected the records from these observers in the hope that he might be able to map the migration routes. A fairly clear pattern emerged. All the ladybirds were migrating either from east to west or from south-east to north-west across the Pennines, choosing routes which took them across the lower ground between the hills from the east to the west side.

The seven-spot ladybird (*Coccinella septempunctata*) is common and widespread from Britain across Europe as far east as India. Swarms often come into England from the continent. One huge swarm in 1952 came in across a 40 mile stretch of the Lincolnshire coast. The high-tide line was marked by a conspicuous pink band formed from the bodies of countless millions which had come down in the sea and been washed ashore. In the 1870s Henri Fabre, the great French naturalist, described a visit he paid to the chapel on the summit of Mont Ventaux, 6,000ft above sea level. The walls and roof of the chapel were so covered with countless thousands of seven-spot ladybirds that from a little distance away the whole building looked as though it might have been decorated with coral beads. They were probably engaged in hibernating or aestivating.

Great swarms of the eleven-spot ladybird *(Coccinella undecimpunctata)* have also occurred from time to time not only in Britain but in many parts of Europe and north Africa. One of the most spectacular was observed by Professor Oliver on 26 April 1939 on the north coast of Egypt. Countless millions were flying in from the sea, and enormous numbers of others had failed to make land and had fallen into the water, to be washed ashore. After the migration had passed there was a continuous drift-line of ladybird bodies at the edge of the water. It was about 1ft wide, 4 to 5in deep, and stretched along at least 14 miles of coast to the west of Alexandria. Professor Oliver estimated that in every foot of this drift-line there was something like 70,000 ladybirds, so that the total number washed ashore would be about 5,000 million.

The ladybird whose migratory movements are best known is the convergent ladybird of the USA (*Hippodamia convergens*). It is a common species in many parts of North America, but has been closely studied in the coastal plains of California. Here it is a valuable ally in the battle against greenfly on crops and in the fruit orchards, this part of California being a great horticultural and fruit-growing area. In the late summer and early autumn the ladybirds leave the plains and fly 100 miles or so into the hills up to a height of about 5,000ft, and here they hibernate in great masses under stones and fallen leaves, on trees, and even on the bare hillsides. In spring they wake up and fly the return journey back to their source of food, the greenfly which are just appearing on the young crops and the fruit trees.

In the early years of this century American entomologists hit upon a bright idea — or so they thought. If they collected hibernating ladybirds from the mountains during winter and kept them in cold storage, releasing them in the fields and orchards on the coastal plain in early spring, they could increase the numbers of ladybirds available to deal with the spring influx of greenfly on the crops. Unfortunately these experiments in biological control did not have the success expected, because within a few days of their release scarcely a ladybird was to be seen. This series of experiments culminated in 1912, when no fewer than 40 million ladybirds were collected, of which it was confidently predicted that each female would lay at least 500 eggs, and would herself eat up to 60 aphids each day. But, as in previous years, they all vanished within days.

Why these releases failed was demonstrated by experiments carried out in 1919. In February of that year 400,000 hibernating ladybirds were collected from the mountains and sprayed with gold or silver paint before being released. Over the next three weeks only 19 were recovered. Towards the end of March a further 600,000 were collected and sprayed before releasing. During the following fortnight only 2 were recovered, one 3 and the other 4 miles away from the point of release. The explanation of these disappearances became obvious. The convergent ladybirds of California have an inborn or instinctive urge to fly 100 miles or so back to the areas where they were reared as soon as they wake up in the spring. If they have been collected and taken back while still hibernating, they still have the same urge when they wake up. They are unable to understand that there is now no need for them to migrate, as they have already been carried to their destination. They therefore set off on their instinctive migration flight, which of course carries them 100 miles away from where they really want to be.

Dr C. B. Williams describes an interesting example of ladybird hibernation which occurred at Rothamsted Experimental Station at Harpenden in England, at which for many years he was the chief entomologist. On 6 November 1935 thousands of sixteen-spot ladybirds (*Micraspis sedecimpunctata*), massed on the south side of a gate post and on the nearby wire fence and the surrounding grass. On the other gate post there were only six ladybirds. Dr Williams thought this was only a temporary resting site, and that they would soon fly off to a more sheltered site for permanent hibernation. Instead they remained there, exposed to all weathers, throughout the winter. On warm days the following April a few woke up and flew off, but the majority remained until 6 May, which was a very warm day, the temperature reaching 72° F. By the end of the day they had all flown off.

CHAPTER 6

NAVIGATION AND ORIENTATION IN INSECTS AND OTHER INVERTEBRATES

Investigations over the past few decades have revealed that some insects at least possess a well-nigh incredible navigational ability. Man has three possible methods by which he can determine the correct direction in which he should go, whether by land, sea or air. From time immemorial he has studied the position of the stars in the night sky, and has used the star map to set a correct course at night. Later he discovered the earth's magnetic field and designed the magnetic compass, which enabled him to navigate both by day and by night with equal accuracy. Both the star pattern and the earth's magnetic field are subject to gradual slight changes, but he soon learnt to make allowances for these variations. His third aid to navigation is the sun, but navigation by the sun demands also the use of a clock, because during the day the sun moves across the sky from east to west. So to determine his direction by the sun he must know what time it is.

COMPLICATIONS OF NAVIGATION BY THE SUN

Despite the complications, it is the position of the sun which insects and certain other invertebrate animals use to determine which way they should fly or walk to reach a chosen destination, or to return home after making a food-hunting journey. Their ability to navigate thus with great accuracy is almost unbelievable. Man can only achieve a similar accuracy by using complicated instruments.

To give some idea of what is involved, let us consider a honeybee leaving its hive to search for a new source of nectar, which means a batch of plants just coming into flower. Yesterday there were no flowers on these plants, but today the first flowers have opened. Before our bee finds them it will probably have flown a considerable distance, maybe several hundred metres, not along a straight path but on exploratory zig-zag courses, sometimes in one direction, sometimes in another, and the distances travelled along these courses will vary. Finally it discovers the new batch of flowers, and after collecting some of the nectar it sets off to fly back to its hive. And, despite the roundabout outward journey, it is able to set a course which will take it unerringly back to the hive along a straight line. Unless it possesses magnetic sense, which we do not know at present, the only way in which the bee can know where it is in relation to the hive is for it in some way to memorise the direction and the length of each part of the outward path. It must then be able to compute all this information and decide in which direction the hive lies. Since it has no magnetic compass it must measure all the various directions by reference to the position of the sun. But the sun is moving across the sky all the time, so the insect has to relate the position of the sun to the time of day to get its bearings, which means that it must also be able to measure the passage of time. It all sounds incredibly complicated, and it is. Man could only estimate his final position after a comparable zig-zag journey in featureless country either by taking measurements of the direction and length of each leg of the journey and then doing a series of mathematical calculations, or by using a computer. But the minute brain of the bee seems to be able to do the job automatically and there is a great deal of experimental evidence that insects and other invertebrates do in fact possess these abilities. First, therefore, let us see what evidence there is to justify the belief that such animals possess an internal clock which they use to guide their normal sequence of activities.

Cockroaches are nocturnal beetles which are often found in kitchens, bakehouses and other warm places where they are likely to find scraps of food dropped on the floor. Normally they sleep during the day hidden away behind or beneath any cover they can find, coming out to feed at night when all lights have been switched off. By excluding all daylight from a room which they inhabited and illuminating it with powerful lights at night it was found that they gradually adjusted their internal clocks to the new conditions. To begin with they continued to come out at night, despite the unaccustomed illumination, and slept through the artificial darkness during the day. Eventually, however, they made a complete adjustment, so that they then slept during the true night, when the room was artificially illuminated, and were active during the day, when the room was blacked out.

Further experiments showed that the cockroach's 'clock' consisted of a group of four cells at the base of the brain. When these four cells were removed from the brain of one of the adjusted cockroaches and replaced by the corresponding four cells removed from a normal cockroach, the behaviour of the adjusted cockroach reverted to normal. Once more it became active at night, when the room was illuminated, and went into hiding during the day, despite the fact that the room was now dark.

Flowers tend to produce maximum amounts of nectar only at certain times of the day, and bees learn to visit these flowers only at these times. Night-flying moths also recognise that certain flowers produce nectar only at certain times during the night, and they, too, are able to confine their visits to these times. In a series of experiments bees from a Paris hive were put into a dark box, flown across the Atlantic, and released in New York. To begin with they continued to forage on a Paris time schedule, which meant that they tended to visit flowers at the wrong time of the day. Soon, however, they were able to adjust their internal clocks so that they worked in New York time. Exactly similar results were obtained when New York bees were boxed and flown to Paris. Thus not only do at least some animals possess accurate internal clocks, but it is possible if circumstances change for these clocks to be reset.

A particularly remarkable example of an activity which is controlled by an internal clock is the emergence of the adult fruit fly, *Drosophila*, from the pupa. Under normal circumstances emergence takes place about an hour before dawn. At this time of day the air is cool and reasonably moist, enabling the soft skin to harden to a protective coat against desiccation before the sun rises. Later emergence during the day would probably result in excessive loss of moisture before the skin had hardened, and consequently the death of the insect. Evidently what happens is that the appearance of dawn on the last few days of pupal development enables the pupae to set their internal clocks. If they are kept in continual darkness they are unable to make this setting, and in consequence they emerge at all hours of the day and night. The clock-setting mechanism, however, must be very sensitive. If at least three days before they are due to emerge the pupae are subjected to just a few flashes of light around dawn, and then the darkness is resumed, this is sufficient for them to set their clocks. When the due day for emergence arrives, although still in darkness, the adult flies will all break out of their pupal cases about one hour before dawn.

Certain animals are able to determine the direction of their movements by using an inbuilt sun compass. Most people will be familiar with the little crustaceans known as sandhoppers, which are common on sandy shores. During rough weather they may be either washed up to the top of the shore and thrown on to dry sand by a breaking wave, or swept out to sea as the water recedes. Those stranded at the top of the beach begin hopping down the beach immediately, the aim being to regain their proper habitat on the middle and lower parts of the shore, while those washed out to sea begin swimming vigorously for the land.

It was thought originally that these return movements were achieved because the little creatures could see what had happened and therefore moved away from or towards the land as appropriate. This was disproved by a series of experiments conducted by two Italian zoologists, L. Pardi and F. Papi. They collected some sandhoppers from the west coast of Italy and carried them across to the Adriatic coast. Some were released at the top of the shore, the others in the water just offshore. On their native west coast the former would have hopped down the shore, that is in a general westerly direction, while the latter would have swum towards the shore in an easterly direction. Despite the fact that, on the Adriatic coast, land was now to the west of them and water to the east, their reactions were precisely the same as they would have been on their native west coast, so that those released at the top of the shore hopped in a westerly direction, which took them inland, while those released in the sea swam in an easterly direction, which of course took them farther out to sea. Both groups perished, one by drying up and the other by drowning. Clearly they fixed the direction of their movements by using their sun compass and not by using any visual clues. In a repeat experiment a batch of sandhoppers from the east coast were transported to the west coast with precisely similar results.

For an animal to be able to navigate by the sun one would imagine that it would need to be able to see the sun. But it has been discovered that insects and crustacea are able to navigate accurately by the sun even though it is obscured by cloud, provided that there is at least some blue sky visible. This need be only one small patch in an otherwise completely cloudy sky, and it doesn't matter where it is in relation to the position of the sun. Only if the whole sky is completely covered with cloud do they lose their ability to navigate. Obviously in some mysterious way they are able to calculate the sun's true position simply by looking at the patch of blue sky.

But how do they do it? The answer is a rather complicated one, and involves reference to a physical phenomenon known as the polarisation of light. Waves coming on to a shore vibrate only in a vertical plane at right angles to the direction in which they are travelling, but light waves from the sun are vibrating in all planes at right angles to their direction of travel. Reflected sunlight, however, is restricted to vibration in only one plane, like water waves, and is therefore said to be polarised. Sunlight reflected through a clear patch of sky away from the sun is polarised because it is being reflected, and the direction of the plane of polarisation depends upon where the blue patch is in relation to the position of the sun. The compound eye of the insect and the crustacean is able to detect the angle of polari-

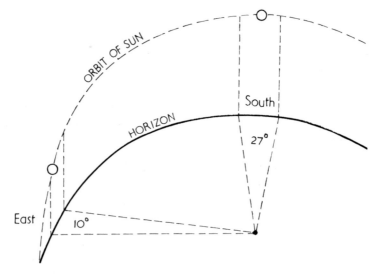

The sun's azimuth or horizontal movement changes less quickly in the morning and evening, when it is rising or setting, than it does in the middle part of the day. The diagram shows the difference in the horizontal angles while the sun travels through the same angle

sation, and their brain is then able to work out from this the true position of the sun.

An additional complication for any animal using the sun compass is that the sun's horizontal movement or azimuth varies at different times of the day. In the middle of the day the sun moves westwards at the rate of 1° in something over two minutes, whereas in the early morning and the late afternoon it will take between five and six minutes to traverse the same horizontal angle. And even these rates vary according to the time of the year.

VON FRISCH'S STUDY OF THE HONEYBEE

The insect whose navigational abilities we know most about is the honeybee, thanks to the remarkable researches of the German zoologist Professor Karl von Frisch, who devoted much of his life's work to the study of this insect. In his early work he investigated the senses of bees, but his most important contribution to knowledge was the discovery and interpretation of the bee dances, by means of which a bee returning to the hive is able to indicate the distance and direction of a source of nectar or pollen which it has discovered. Beekeepers had always known that foraging bees were able to do this, but it was von Frisch who showed how they were able to do it.

Although one or two authorities had suggested that bees could see colours it had always been generally accepted that bees were colour blind. Von Frisch's first important researches showed that they did possess colour vision, but that it was different from our own. When we look at a rainbow we see a continuous series of colours across the width of the rainbow from the red at one edge through orange, yellow, green and blue to violet at the opposite edge. To the bee, however, the rainbow appears as a series of coloured bands separated by grey bands, these representing the colours it cannot see. It cannot for example see red, so the red band appears black and the orange band dark grey. The colours it can see are yellow, blue-green and violet, each of these being separated by bands of grey. In addition, however, the bee is able to see another coloured band beyond the violet. This is the ultra-violet band, to which our eyes are insensitive. The ability to see ultra-violet has an important bearing on the colours of certain flowers as seen by bees. Many red flowers appear black but some others, for example red poppies, appear coloured, because in addition to red they also reflect ultra-violet light, which we cannot see but the bee can. The various kinds of white flowers appear different to bees, depending upon the amount of ultra-violet light they reflect.

With this knowledge von Frisch was able to devise a method of observing what went on within the darkness of the hive without the bees realising that they were being watched. He inserted a red-glass window in the side of the hive facing a vertical comb. So far as the bees were concerned the window let in no light that they could detect, so they were able to continue their activities in what to them was normal darkness.

What he observed surprised him. He found that bees returning to the hive indulged in a kind of dance on the vertical comb, while other bees which had remained in the hive crowded round as though watching the dance, though

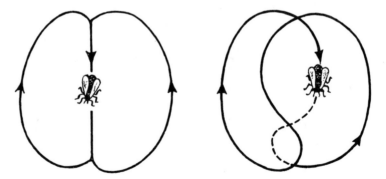

(left) The bee's round dance; (right) the tail-wagging or figure-of-eight dance

they couldn't of course actually see anything in the darkness. There were, too, two quite distinct types of dance; a round dance in which the bee completed a circle in one direction and then reversed and completed one in the opposite direction, often continuing to alternate in this way for a considerable time; and a more complicated tail-wagging dance. This took the form of a figure-of-eight, consisting of a left and a right circle with a straight run between them. Whenever the bee was travelling along this straight section it was seen to wag its abdomen or tail end from side to side — hence the name which von Frisch gave to this dance.

Von Frisch's first interpretation of these dances was that the round dance indicated that the returning bee had found a new source of nectar, and the tail-wagging dance that it had located a source of pollen. In order to verify his theories he devised a series of investigations which not only disproved his original suppositions, but revealed a complex and very sophisticated system of communication.

The method he used was to place dishes of sugar-water at various distances and different directions from the hive. Each bee visiting a dish to collect sugar-water was marked by a distinctive blob or blobs of paint, so that it could be identified when it returned to the observation hive. His first discovery was that the method of dancing conveyed to the bees in the hive information about the distance away of the source of food. It it was less than 100yds the returning bee danced the round dance, but if it was more than 100yds it executed a figure-of-eight tail-wagging dance. As a further refinement the bee varied the number of figures danced in a given time according to the distance. If this was only a little over 100yds about 20 figures would be danced in a minute, whereas if the source of food was 2 miles from the hive only 4 patterns would be danced in this time. An intermediate number of figures danced indicated intermediate distances.

Although unable to see the dancing bee, those bees crowding around it in the hive were able to sense what it was

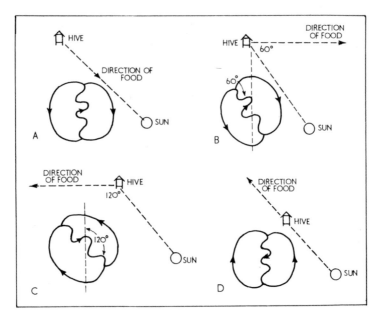

Diagram to show how a bee returning from a foraging expedition uses the tail-wagging dance to indicate the direction in which the other bees should fly to locate the food it has discovered. The direction relative to the current position of the sun is indicated by the position of its body as it traverses the central part of the dance.

In figure A the bee is going straight up, indicating that the food is directly towards the sun. In figure D the food is directly away from the sun, so the bee dances vertically downwards. In figure B the food lies in a direction 60° to the left of the sun, so the bee dances 60° to the left of the vertical. In figure C the bee is indicating a direction of 120° to the right of the sun by dancing at an angle of 120° to the vertical, which of course carries it in a downward direction

doing by maintaining close bodily contact and by examining its body with their antennae or feelers. These were able to pick up the individual scent of the foraging bee's body, which the bees in the hive would recognise on the flowers from which it had returned when they went out to search for them. The antennae could also pick up the scent of the flowers which would have adhered to the returning bee's body, as well as the taste and scent of the nectar or pollen from these flowers, because they would each be given a minute sample by the returning bee.

But much more information than the distance of the source of food from the hive, the scent of the flowers and the scent and taste of their nectar and pollen was given by the dancing bee. It was also able to indicate in which direction the bees should fly from the hive in order to reach the source of nectar of pollen which it had discovered. In the round dance if the bee reversed direction at the top of the circle this indicated that the source of food lay directly towards the sun, and if at the bottom of the circle, directly away from the sun. Directions to the right and to the left of the sun were indicated by reversing direction along the right or left semi-circle, the exact position of the reversal giving the correct angle.

In the tail-wagging dance the correct direction relative to the present position of the sun was indicated as the bee traversed the vertical central part of the dance. If it went straight up, wagging its tail from side to side, then the food was directly towards the sun. If however it went straight down, this indicated that the food was directly opposite the sun. To indicate direction to the right or left of the sun, the bee pointed its body at the correct angle to the right or left of the vertical line as it wagged its way up or down the central part of the dance.

Sometimes a returning bee will go on dancing for a considerable time as it gives navigational instructions to a number of groups of bees. During this time it has to take account of the fact that the sun does not remain still but is moving westward across the sky at an angle of about 1° every 2½ minutes during the middle part of the day. It must therefore use its internal clock to adjust the angle of its dance to compensate for the sun's movement. On one occasion a bee was observed to dance continuously for 84 minutes. During this time the sun would have swung through an angle of 34°, while the bee changed the angle it was indicating by 33°.

One thing a dancing bee cannot do, and that is indicate

height. If it has located a source of nectar or pollen on the other side of a hill, all it can do is to indicate the compass direction. But the bees who have been given this information usually prefer to make a detour round the base of the hill to reach it rather than rise in the air to fly over. How they are able to achieve a semi-circular flight around the base rather than a straight one over the top, we do not know. On one occasion von Frisch placed his hive directly beneath a tall radio mast and hauled a dish of sugar-water to the top. A few bees followed and collected some of the water, but when they returned to the hive they found themselves unable to give their fellows any precise directions. They were excited and they danced, but their dancing was confused, and indicated all kinds of directions.

The dances normally take place on the vertical surface of the comb inside the hive, but sometimes a returning bee dances on the alighting board outside the hive if there are any bees there when it arrives. It then dances in such a way as to point towards the true direction of the discovered food. In order to find out what would happen if the comb was laid horizontally, von Frisch arranged his experimental hive so that it could be turned on to its back. The bees were not in the least put out. Like those which danced on the alighting board, they now indicated the true direction.

We have already mentioned the insects' internal clock, so essential if they are to use the sun as a compass with any degree of accuracy. Von Frisch carried out a series of experiments which reinforced the evidence for its existence. His aim was to train bees to come to a dish of sugar-water at certain times of the day. In his first experiment he placed the dish on a table at 4pm and removed it at 6pm (16.00 and 18.00hrs on the 24hr clock), marking each bee which came to feed during this 2hr period, and giving it a reference number. He continued this intermittent feeding for about a week. On the following day the dish was left on the table all day from sunrise to sunset, and an observer was stationed nearby throughout the day to record the bees, with their reference numbers, which came to the now empty dish during each ½hr period. One bee, No 11, appeared once between 7am and 7.30am to explore the dish, and a second time between 7.30am and 8am. No other bees appeared until the half-hour before the training period, between 3.30pm and 4pm, when 2 appearances were made. Then came the training period. Between 4pm and 4.30pm bee No 1 made 4 appearances and bee No 12 made 2 appearances. In the next half-hour up to 5pm there were altogether 17 appearances, bee No 17 returning no fewer than 10 times in a vain search for food. From 5pm to 5.30pm there were 11 appearances, and in the last half-hour of the training period only 4 appearances. From 6pm to 6.30pm there were 2 appearances. Thus out of a total of 44 appearances only 6 were outside the training period. Clearly bees were skilled at estimating the time of day on their internal clocks. In further experiments von Frisch was able to train bees to come to the feeding table at several times during the day.

To eliminate the possibility that these trained bees were using the sun or the occurrence of day and night rather than an internal clock to estimate the correct time to appear at the feeding table, von Frisch next carried out a series of experiments in which the hive was transferred to a room from which all daylight was excluded but which was permanently illuminated by electric lights. He found that it was just as easy to train the bees to come for food at a given time in the twenty-four hours as it had been to train them in a hive exposed to day and night. It was possible, too, in the constantly illuminated room, to train them to come for food at any time during the night. He was also able to show that they worked on a 24hr clock: when he fed them every 19 hours they proved incapable of learning the rhythm, even over a long period of time, when he tried to train them to come for food every 48hrs they appeared and searched for food every 24 hours despite all his efforts.

NAVIGATION IN ANTS

Another insect whose navigational abilities we know a good deal about is the ant. Unlike the bee it does not fly, apart from the brief nuptial flights undertaken by males and queens at the breeding season. But it does possess the same kinds of abilities as the bee.

Before World War I an Italian zoologist, F. Sanschi, conducted a series of experiments on north African ants. He surrounded foraging ants with a high screen so that they could not see any familiar landmarks which might be of use to guide them back to their nest. He also held an opaque disc over them so that they could not see the sun. Yet they were still able to set an accurate course in the direction of the nest. Sanschi suggested that they were able to see the stars and navigate by them, just as we are able to do even in daylight if we look up from the bottom of a well. In fact, as we now know, ants, like bees, possess a sun compass and are able to deduce the position of the sun, even if it is hidden by cloud, by using the polarised light from a patch of blue sky however far away from the sun it may be. If an ant is picked up and removed to the right or the left as it is making for its nest it will continue to walk in the same compass direction, even if it is turned to face the opposite way when it is put down. This will of course result in it missing its nest by the same distance as it was displaced.

Ants also possess an internal clock, and are able to make allowances for the changes in the sun's azimuth as it travels across the sky. In June 1964 Dr J. Reimann, a German zoologist, collected a group of 58 ants which were making for their nest at 11am, and put them into a light-proof box. Their direction at the time of capture was at an angle of 90° to the right of the sun. At 2pm he released them at the same spot, and they set off in a direction only 6° to the right of the sun. They had clearly calculated that the sun's azimuth had shifted 84° westward during their 3hr imprisonment in the dark. In fact it had shifted 86°, but the 2° error was so slight that they would still arrive safely at their nest.

We have already mentioned that the rate at which the sun's azimuth changes depends upon the time of day. To find out whether his ants were capable of adjusting to these changes Dr Reimann took a second group of 47 ants from the same nest at 4pm, shut them away in darkness for 3 hours and released them at 7pm. During this second 3hr period the sun's azimuth had moved through only 34°. The ants changed their direction with respect to the sun's direction by 39°, again sufficiently close to see them safely home.

Navigation using the sun, polarised light and an internal clock is not the only method used by ants. A foraging ant discovering a new source of food will gather as much as it can cram into its crop and set off for the nest, probably using sun navigation to fix its course. But as it goes it lays a scent trail which other ants will be able to follow. When it arrives back in the nest other ants gather excitedly around and pick up both the scent of the food and of the returning ant. They then leave the nest and follow the trail themselves, each one reinforcing the scent so that it gets stronger. If a human finger, which smells very strongly to an ant, is rubbed across one of these ant trails it will wipe out the ant scent at this point. This causes temporary confusion among the next ants to come along the trail. They will cast about for a time until one ant discovers the continuation of the trail beyond the break, and the trail will then soon be restored as other ants follow.

SCENTING ABILITY OF MOTHS

Scent is used, too, by other animals. Perhaps the most astounding scenting ability is exhibited by the males of certain species of moths in detecting and locating newly-

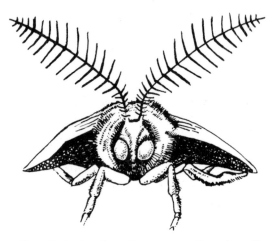

The well-developed sensitive antennae of a male moth

emerged females. Most of the research work in this field has been carried out on the emperor moth, but the findings have been verified in experiments with other species. When she first emerges from her cocoon a female moth is not ready to mate, and is therefore odourless. After a few hours, however, she is in the right condition for mating, and she produces, in two small special pouches at the end of her abdomen, a perfume designed to attract male moths. In order to disseminate this perfume into the air she raises her abdomen and opens out the pouches; only a few molecules of perfume landing on the antennae of a male sets him off to seek the female. So effective is her signal that she can attract males from as far away as 3 miles in a relatively short time. Experiments have shown that a male moth 1 mile away from the female can be with her 10 minutes after receiving the signal, representing a flying speed of 6mph, considerably faster than a man can walk. The male antennae also act as direction finders. The male will orientate himself in flight so that the number of female scent molecules reaching his right antenna equals the number being received by the left. In this way he will home accurately on the female source of the disseminated molecules. In experiments the two antennae of a male moth have been fixed across each other with a piece of delicate thread. The messages received by the moth have then been so confused that he has been unable to locate the female.

LOCATION BY RADIANT HEAT

In Chapter 4 we saw how rattlesnakes use radiant heat to detect their prey. Many insects also use the heat radiated from the bodies of their victims in order to locate them. Female mosquitoes are in this way guided to the bodies of the particular mammal species whose blood they must suck up before they can lay a batch of eggs, the blood providing them with the necessary protein which must be stored in the eggs. As with the scent detected by male moths, it is the antennae of the female mosquito which are used to locate and home on the source of heat. Fleas which have just hatched are sensitive to the heat radiated from the bodies of their potential victims. Ticks, too, are able to detect their victims because of the radiant heat they produce. There is a similar explanation for the well-known ability of bed bugs to crawl over the ceiling at night and drop down on to any part of the human body exposed above the bedclothes. They use their antennae to detect the heat which is being radiated by the exposed limb, and then simply let go.

ORIENTATION BY LIVING LIGHT

Some insects are able to orientate by means of living light which is produced in special organs. The best-known of these luminous insects are the glowworms and fireflies. Despite their name glowworms are really beetles, various species of which are common in Europe and in other parts of the world. In the European species only the males are able to fly, the females being wingless. But their lights are much brighter than those of the males. During the breeding season they sit in conspicuous places on hedgerows and bushes at night emitting a steady glow, which the males can pick out from some distance away as they fly around looking for a mate. As soon as a male spots a female's guiding light he flies in to settle beside and mate with her. The light is a greenish colour, and is produced in a number of light organs situated on the abdomen. Unlike light produced by electricity or burning oil or gas, whose production is accompanied by a good deal of heat, living light is cold light unaccompanied by heat. The substance which is actually responsible for producing the light is luciferin.

Fireflies are mainly confined to tropical and subtropical regions. They are related to the glowworms, but their signalling system is more complex and more spectacular, though it also enables the males and females to find each other at night. Instead of producing a steady light they signal by means of a number of flashes separated by periods of darkness. Each species has its own particular code, so that a male of one species will not normally be attracted to a female of another kind. There is one interesting exception, however. Just as Cornish wreckers lured ships on to dangerous rocks by extinguishing genuine warning lights and replacing them with lights placed on cliff tops, so the females of one species, *Photuris*, are able to entice the males of another species, *Photinus,* by copying the signal which *Photinus* females use to lure their own males. As soon as an unsuspecting *Photinus* male lands beside a *Photuris* female he is pounced upon and eaten. The *Photuris* of course also has her own code for enticing males of her own species to fly in to mate with her.

CHAPTER 7

MIGRATION AND NAVIGATION
IN BIRDS

Migration is more widespread among birds than among any other group of animals. There are several possible reasons for this, one being that it is much easier for any animal which flies to undertake long journeys with the minimum of effort than it is for animals which live in water or on land. In the northern hemisphere the most obvious explanation of migration would appear to be that birds fly towards the equator before winter sets in to avoid severe weather. But the mere avoidance of cold is not the real answer. After all, birds are warm-blooded and well supplied with an insulated layer of feathers which would certainly enable them to withstand the coldest weather conditions.

The true reason for most migrations is however connected with the onset of cold weather, not because the cold weather itself would have a detrimental affect on the birds, but because of its effect on their food supply. Seed-eating birds can find plenty of food whatever the weather, but birds which feed on insects would find little to eat during the winter, which explains why swallows, cuckoos and other insectivorous birds must fly to warmer climates by then to find adequate food supplies. The position with sea birds is similar. During the months of spring and summer the Arctic Ocean and the adjoining land provide an abundant supply of food, but virtually none in the winter, and the same is true to a lesser extent somewhat farther south in the temperate zone.

All this is probably fairly obvious. But what is not so

apparent is why birds should forsake the tropical and subtropical regions for the summer months. Why don't they stay where there is plenty of food all the year round? The reason is that although well north or well south of the equator winter days are very short, giving little time for feeding even if there was plenty of food available, the converse is also true. During the summer months the day is much longer in temperate and arctic and anarctic regions than it is in tropical regions. So if the birds migrate to subpolar or temperate regions for the summer they obtain many more hours of daylight for collecting food to feed themselves and their young than they would in regions nearer the equator, and food at this time of the year is more abundant. Some birds make the best of both worlds by living in the north temperate or even arctic zone when it is summer there, and migrating to the far south to the southern temperate or antarctic zone when the northern winter sets in.

Research over the past few decades has thrown a great deal of light on the whole problem of bird migration. Birds, we now know, use every possible method of navigation. Some are able to navigate by the sun and polarised light, using their internal biological clocks; others are able to navigate at night by using the stars. They are also able to navigate by the more normal method of memorising landmarks, sometimes known as terrestrial navigation.

MIGRATION JOURNEYS OF SEA BIRDS

One of the first authorities to investigate bird migration and navigation and provide us with indisputable evidence that birds do in fact cover vast distances and possess astonishing homing abilities was R. M. Lockley, the British ornithologist who has devoted much of his life to a study of bird migration and is now one of the world's leading authorities on the subject. His first classic researches were conducted on the Manx shearwater (*Puffinus puffinus*), which nests in its

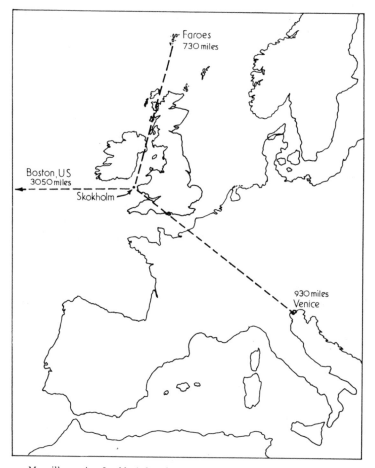

Map illustrating Lockley's homing experiments with Manx shearwaters

thousands on his island home of Skokholm off the west coast of Wales. But although the Manx shearwater nests in Europe, mainly on islands around the British Isles as well as Ireland and Scandinavia, it spends the winter in the waters off the east coast of Brazil, some 6,000 miles away. They return to Europe in late February and March, and begin nesting in late April in rabbit burrows on the cliffs; living and nesting in these burrows to avoid the various gulls, ravens, crows and other birds of prey.

Lockley first established that the nesting birds take it in turn both to incubate the one egg and to look after the single chick which hatches from it. The bird which is off duty is away for up to six days, so he decided that if he caught some of the birds just as they were being relieved by their mates he could conduct homing experiments with these birds without endangering the egg or the chick. Adult birds thus taken from the nesting burrows were put into closed boxes after having been ringed and were sent by air to various places to be released. All these places were well away from their normal migration route between Skokholm and South America. They were in fact released from rooftops in Cambridge and Birmingham, from slopes in the Swiss Alps, from the side of a lagoon in Venice, from a ship in the Faeroe Islands and from the edge of Boston Airport. Those releasing the birds were asked to note the weather conditions at the time of release.

One of the two birds released at Venice was lost to view almost immediately, but the other was kept in view long enough to establish that it orientated itself rapidly and set off in a direction which would lead it straight to Skokholm. The interesting point about this is that although the Adriatic Sea was visible to the south-east from the point of release the shearwater — a completely aquatic bird which will not normally fly over land except the few yards necessary to achieve its cliff-top burrow — ignored the sea completely, showing complete confidence in the overland direction it had chosen. All the other birds, too, seemed able to set an accurate course for Skokholm. Perhaps the most remarkable was the single bird released at Boston Airport, which is at the edge of the sea. This bird crossed a little more than 3,000 miles of ocean in 12½ days. Allowing for the fact that it would have to rest and to feed for probably 12 out of each 24 hours, its actual flight speed for the journey would be about 20 miles an hour, which is known to be its normal speed. There would have been no time therefore for it to search the whole of the Atlantic with the hope of eventually finding its nesting site; this could have taken months, and the chances of success would have been minimal. Clearly these released birds were able to estimate the direction of home very accurately. One very interesting fact revealed by these experiments was that provided the sun was out or there was at least some blue sky, the birds were able to choose the correct direction of flight almost immediately, but if the whole sky was overcast they tended to fly around for some time, as though uncertain which way to go. Those birds released under these conditions took much longer to return to Skokholm, and some in fact never returned at all. This strongly suggests that shearwaters use celestial navigation just as bees do.

Although normally they fly by day, shearwaters only return to their breeding burrows at night, both when arriving in the early spring from South America and after their feeding sorties during the incubation of the eggs and the brooding of the chicks. As explained in Chapter 3 this change in normal habits is designed to avoid the attentions of predators, and since on moonlight nights the birds home silently but on dark nights come in screaming, they apparently use a form of echo location.

The shearwaters which appear from South America early in the season find little to eat. Indeed until midsummer the seas off the Welsh coast contain very few small fish on which the shearwaters can feed. By ringing experiments Lockley was able to establish where these first arrivals and the parents leaving their nests to feed until midsummer found their food. Early in the year there are large numbers of sardines, hatched the previous year, swimming in the Bay of Biscay. These small fish gradually swim northward. In late February and March the shearwaters have to fly all the way from Skokholm to Biscay to feed. In April and May, as the sardine shoals move farther north, the birds have a shorter distance to fly. Eventually, by about the middle of June, the birds no longer have to fly considerable distances in search of sardines, because by this time there are plenty of tiny fish in Welsh waters.

About sixty days after hatching the young shearwater is at last deserted by its parents. It has been well fed during the time that they have looked after it, and for the next week or so it will fast. But at night it will come out of the now unattended burrow in order to exercise its wings. Suddenly it, too, will forsake the burrow in which it was raised and, under cover of darkness, glide down from the cliff into the sea. Within a short time it will become airborne and set off, without visible guidance but obviously with some inherited knowledge, to follow its parents to the winter quarters of the species 6,000 miles away in South America. The purpose of these migrations from one hemisphere to another is to enable the birds to live under summer conditions all the year round, with food much more abundant than it would be in the winter.

The Manx shearwater is not the only member of the family to undertake spectacular journeys. The slender-billed or short-tailed shearwater (*Puffinus tenuirostris*) of the southern hemisphere, was first discovered in 1798 when Captain Flinders sailed his ship through the strait between Australia and Tasmania, and named it Bass Strait, after his ship's surgeon, Dr Bass. They recorded a bird which 'passed over us continuously for 90 minutes in a dense flock about 300 yards wide. We estimated their number at 151 million'. The flesh of the birds is said to taste like mutton, which explains the name 'mutton bird' by which they are also known. In the early 1960s two Australian zoologists set out to find where these enormous flocks came from and where they nested. In all they ringed 32,000 birds, and were in consequence able each year to establish their whole life-history. What they discovered was that the birds arrive first on a number of islands between Australia and Tasmania during the nights of 26 and 27 September. After a few days of courting and mating they disappear as suddenly as they arrived, presumably to feed in the open ocean. Then, just as suddenly, they reappear on 19 November. This time the females lay their eggs in hollows in the ground. At first the males incubate the eggs for about a fortnight while the females are away fishing on the high seas. On their return the males go off on a fishing trip of similar duration. This alternation goes on until the chicks hatch in January and continues throughout their feeding period which lasts until the middle of April. Then begins the great annual flight. First they fly some 6,000 miles northward to the Sea of Japan. By June they have reached the Bering Sea in the far north between Alaska and Siberia. But still they move onward, though now flying south-east, so that by August they are off the Pacific coast of Canada. When they finally reach the latitude which divides Canada from the United States they suddenly veer to the south-west then across the Pacific Ocean via the Hawaiian and Fiji Islands to return to the islands of the Bass Strait on 26 and 27 September. Since leaving some five months earlier they have flown about 20,000 miles!

Another champion migrator among sea birds is the Arctic tern (*Sterna paradisaea*), whose annual journeys are longer than those of any other bird. They nest in the Arctic during the Arctic summer then, as autumn approaches, they leave their nesting sites and fly south across the equator until they reach the Antarctic, where they enjoy the excellent feeding available during the Antarctic summer. When the time to migrate southward arrives, those birds which have nested on the Canadian islands of the north Atlantic fly eastwards to the southern tip of Greenland, where they join up with those birds which have nested along the Greenland coast. Together they continue to fly eastward until they link up with birds which have nested in Iceland. The whole flock

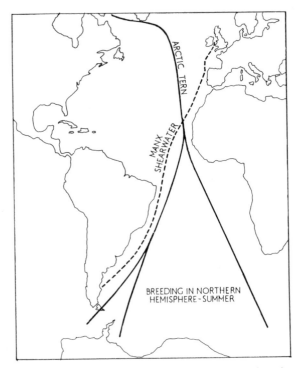

A number of sea birds travel enormous distances twice a year in order to live in perpetual spring and summer. Some, notably the Manx shearwater and the Arctic tern, breed in the summer of the northern hemisphere and fly south to spend the southern summer in the southern hemisphere

now veer southward, making for the west coast of Africa. Off Dakar the flock divides into two: half continue to fly southward past Cape Town until they reach the Antarctic continent; the other contingent turn westward across the Atlantic until they reach Brazil, where they turn southward. Some then continue to follow the coastline to reach the Antarctic via Tierra del Fuego, while the remainder get there via the Falkland Islands. Meantime those birds which have nested in Alaska have been flying southward along the west coast of North and South America. These finally join up with those which have arrived via Brazil. On these annual return journeys they cover anything up to 25,000 miles.

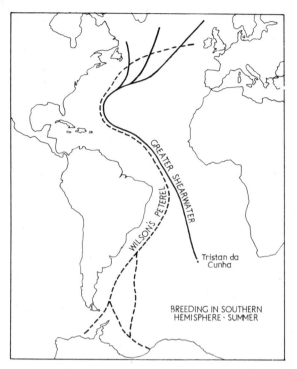

In contrast to the Manx shearwater and the Arctic tern, which nest in the northern hemisphere, the greater shearwater and Wilson's petrel nest in the southern hemisphere. But they too make enormous migration journeys

Hardly less remarkable are the migrations of the greater shearwater (*Puffinus gravis*). This bird does not occur in the Pacific, but during the north Atlantic summer it lives and feeds over a wide area from Newfoundland across Greenland, Iceland and Scandinavia. By some miracle of navigation about 4 million birds are able to set out from their various feeding sites in the autumn and fly unerringly across between 7,000 and 8,000 miles of completely featureless ocean to arrive at the tiny Tristan da Cunha group of islands, where they nest.

Another bird which feeds in the northern hemisphere and nests in the southern is Wilson's petrel (*Oceanites oceanicus*). It nests during the Antarctic summer on the

123

edge of the Antarctic continent, and then flies north to spend the Arctic summer as far north as Newfoundland.

Albatrosses have always been known as great travellers. We now know that the large wandering albatross (*Diamedea exulans*) flies continuously round the world in a westerly direction in the region of latitude 40° south. It takes advantage of the almost continuous westerly winds and gales which occur in this region—the 'roaring forties'. At this latitude the complete distance round the world is about 20,000 miles. The birds pause only to nest and rear their young at the appropriate time on remote islands.

Scientists of the United States navy made an interesting discovery about the homing ability of another species, the Laysan albatross (*Diomedea immutabilis*). When the navy wanted to build an air base on one of the Midway Islands they found a colony of these birds nesting in the area where they intended to construct the runway. Not wishing to harm the birds the scientists experimented with the possibility of catching them and releasing them elsewhere. In a trial run they collected eighteen and transported them by air to various places in the Pacific 3,000 or more miles away. The experiment failed. Within a short time fourteen of the albatrosses had returned, one of them only ten days after its release.

MIGRATION JOURNEYS OF LAND BIRDS

One might perhaps expect sea birds to be able to travel vast distances over the oceans. After all except during the breeding season they spend their whole time at sea, from which they also derive all their food. Much more remarkable are the migration journeys which various land birds are able to make across great distances of sea and ocean, because they would be unable to swim if they were forced down on to the water. One of the most remarkable of these journeys is undertaken by a humming bird weighing only ⅛oz. During summer it lives and breeds in the southern United States, but with the approach of autumn and colder weather it realises the need to find a warmer climate for the winter. To achieve this it sets off across the Gulf of Mexico, a prodigious journey for so small a bird of about 500 miles. Since it cannot swim it has to keep going, and it covers the distance in about 10 hours, thus achieving an average speed of 50 miles an hour. For some time before setting out it feeds hard to lay down a store of fat in its minute body which will provide it with the necessary amount of energy for the journey. In the following spring it undertakes the return journey with equal success.

Another land bird which undertakes a long sea journey twice a year is the New Zealand bronze cuckoo (*Chalcites cuccolus*). It breeds in New Zealand and, like its European counterpart, lays its eggs in the nests of other birds, which then become responsible for bringing up the nestlings. As soon as the females have deposited their eggs in the foster parents' nest both they and the males set off on the long journey back to their winter quarters. First they cross the 1,250 miles of sea separating New Zealand from Australia. After a pause of a few days to rest and feed they are off again. They now fly north over the water until they reach New Guinea, finally turning north-east to reach a group of small islands off New Guinea's north-east coast known as the Bismarck Archipelago. The total journey from New Zealand covers nearly 4,000 miles. It is certainly a feat of navigation for the adult cuckoos to reach their destination safely; but how much more so for the young cuckoos. By the time they are fledged and are able to leave the foster parents' nests their parents have vanished, and cannot therefore guide them on their very first journey to their winter quarters. Nevertheless they have some inborn guidance system which unhesitatingly enables them to cover the 4,000 mile journey to join their parents.

Even more remarkable, if possible, is the journey under-taken by another land bird which cannot swim, the American golden plover (*Pluvialis dominica*), a relative of the lapwing. After nesting in the summer in Alaska they set off on a 2,500 mile ocean flight to their winter quarters in Hawaii. Some of them are apparently not content with the Hawaiian climate for, after a short rest, they go on to the Marquesas Islands, making another 2,500 mile ocean flight.

Equally astonishing are the return migrations of the Mongolian plover. This bird nests in Siberia, but as autumn approaches some fly southwards through Malaysia and the Indonesian Islands to winter in Australia, while the remainder fly via India and the 3,000 miles of the Indian Ocean to reach South Africa.

Spectacular though these migrations of land birds across the oceans may be, the majority of land birds of course migrate across land, and a study of their journeys has revealed a great deal of interesting information. The Alberta crow spends the summer months in Canada, where it nests; but before the severe winter weather sets in it heads south for the milder climate of Oklahoma, where it spends the winter. Professor William Rowan of Alberta University reared some young crows in captivity. He then waited until the winter snows had arrived, long after the other crows had left on their journey south, before releasing his young captives. With no one to guide them, and with no previous knowledge of the route, they nevertheless flew southward without hesitation until they joined up with the rest of the crow population in Oklahoma.

Experiments with European hooded crows (*Corvus cornix*), also demonstrated the possession of an inborn sense of migration direction. In the spring these birds migrate north-eastward to nest in country bordering the northern Baltic sea. The late Werner Ruppell captured 900 young crows on migration at Rossiten on the German Baltic coast, and ringed them all. Four hundred were released immediately to continue their journey, while the remainder were transported 465 miles due west and released on the borders of Germany and Denmark. Subsequently it was found that these latter birds had continued their journey in the same direction they would normally have taken, and had nested in an area that distance to the west of their normal nesting area, in a region where hooded crows are unknown.

Storks also apparently have an inborn sense of migration direction. In one experiment 144 young birds were collected in East Germany and reared in captivity in West Germany. Now East and West German storks both winter in the Nile valley, but they get there by quite different routes. The western flocks fly southward through France and Spain to Gibraltar. Here they turn eastward, flying all along the coast of North Africa until they reach Egypt. The eastern storks, however, set off in a south-easterly direction which brings them to Greece. Here they turn east, and reach Egypt by finally flying south along the coasts of the eastern Mediterranean countries. The transported storks were kept in West Germany until all the native birds had departed on their autumn migration before being released. They flew off south-eastward without hesitation, subsequently joining up with the eastern flocks in south-eastern Europe. Clearly their direction sense was inborn. This experiment was repeated the following year using 754 transported storks, but these were released just as the native western flocks were beginning to migrate, and the result was very interesting. Instead of flying south-eastward, they joined these western birds and reached the Nile valley via Gibraltar. In the following year some of them even returned with the western birds to breed. In this case the stimulus of other birds of the same species, but not of the same population, migrating, was able to override the inborn direction stimulus.

NAVIGATION BY THE SUN

Cuckoos of course must be provided with an accurate inborn set of migrating directions, for the adult birds, having deposited their eggs in the nests of other birds, have nothing further to do with their offspring. There would therefore be no opportunity for the parents to teach their young anything, and they certainly could not act as guides on migration because they have flown off to Africa long before the young birds are ready to leave. Yet the latter are able to follow their parents and eventually meet up with them in their winter quarters. As long ago as 1935 Bastian Schmidt noticed that hand-reared cuckoos kept in aviaries became restless as the proper season for them to migrate approached, and that when this happened they always sat on their perches facing south, as though pointing in the direction they would fly if released. But at this stage there were no theories as to how cuckoos and other migratory birds might navigate. It was not until after World War II that biologists began seriously to speculate how birds managed to find their way, not only by day but by night as well. Dr G. V. T. Matthews in England and Dr Gustav Krämer and Dr Franz Sauer in Germany, with other workers over a period of about twenty years, were able to show that birds did indeed navigate by the sun during daytime, using polarised light to locate the position of the sun even when it was obscured by cloud provided at least a small patch of blue sky remained visible, and that during the night they were able to navigate by the stars.

One of the first birds to be used in these investigations was the starling. Each autumn huge flocks which have spent the summer in Denmark and the Scandinavian countries fly south-westward through Holland to their winter quarters in northern France and southern England. In 1957 Dr A. C. Perdeck captured 11,000 birds on migration in Holland, ringed and transported them to Switzerland, where they were immediately released. The results were enlightening. Young birds which had been hatched only a few months before possessed a strong inborn sense of the direction they should fly on migration, and these flew south-westward, though this brought them to Spain and Portugal, where northern starlings would normally never be found, instead of to England and France. But the older birds, which had already migrated at least once before, were able to change the direction of their flight to north-west, which brought them to northern France and England.

In an ingeniously designed experiment Dr Kramer was able to show conclusively that starlings were capable of recognising a compass direction provided they were able to see the sun or at least a patch of blue sky. He constructed an aviary for twelve birds with solid revolving walls which shut out all light except that coming from the sky above. This meant of course that the birds could not see any landmarks. In the centre there was a table whose outer edge was also capable of being revolved, and equally spaced around it were twelve identical food boxes. To get at any food which might be placed inside, the birds had to push aside a flap. The experimental bird was placed in a cage on the fixed central part of the table.

For a number of days each experimental bird was trained to find food in one of the boxes in a certain compass direction, perhaps east, perhaps west. Each bird soon learned to look only in the box which was in the right direction and which contained food, and not in any of the other eleven boxes which in fact contained no food. Once conditioned the bird was then ready for the next stage in the experiment. With the bird in its cage the outer edge of the table containing the boxes and the walls of the aviary were both rotated, either in the same or in opposite directions. When they were stopped the cage was lifted, and on every occasion the bird made for the box which was in the compass direction it had been trained to go. Normally it would find food in the box, but if there was no food the bird looked perplexed, but did not try any of the other boxes. Kramer made one extremely important observation. Before the bird walked when the cage was lifted, it cocked an eye

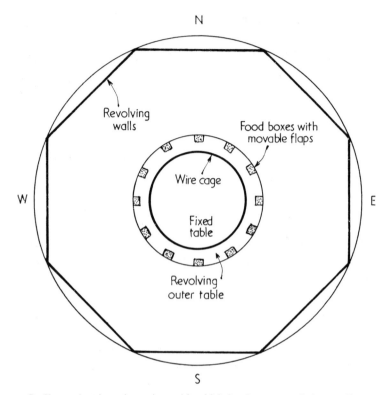

Dr Kramer's orientation aviary with which he demonstrated that starlings could recognise a compass direction even if they were able to see no more than either the sun or a patch of blue sky through the transparent roof of their cage

skywards as though deciding in which direction it should go. If the sun was shining, or if it was covered with cloud but there was at least one patch of blue sky visible, the bird never had any difficulty in moving off in the correct direction. But if the sky was completely overcast it was unable to determine the correct direction, and searched the boxes at random.

Kramer carried out further tests on his birds by transferring the entire apparatus to a room without windows. A single lamp was used to imitate the sun. The birds proved able to orientate themselves accurately so that they went to the correct food box in relation to the position of the false sun even though the false sun was in a quite different position in the 'sky' from that of the true sun.

⌈Many birds of course spend their whole lives in the same area and never migrate. So far as we know they lack the special navigational abilities possessed by migratory birds. If therefore they are caught and released at considerable distances away from their home territories they are unable to return.⌉ In this connection the homing pigeon is particularly interesting. It is descended from the rock dove, a bird which does not migrate, yet it has acquired considerable ability at finding its way home after being released at varying distances of up to 100 miles or more from its home territory. Its homing ability, however, is not comparable with that of true migratory birds such as the Manx shearwater! One fact well-known to pigeon fanciers is that in order to orientate correctly for home the released pigeon must be able to see the sun when it is released. In misty weather many birds never reach their home lofts, and those that do take much longer than normal to complete the journey.⌉

To find out more about the homing pigeon's abilities and limitations Dr Kramer carried out a series of experiments. Young pigeons were raised as nestlings in aviaries at Wilhelmshaven in northern Germany. They were able to fly within the aviaries and could see the sky and the horizon, but were never allowed outside. They therefore had no chance of learning what their aviaries or their surroundings looked like from the air. When fully fledged Kramer took them nearly 100 miles south to Osnabrück for what was to be their first flight. On being released they all succeeded in returning to their home territory, many of them being found perched on the roof of their aviary. Clearly the birds had been able to obtain a good deal of information about the position of their aviaries from inside. But how much, Dr

Kramer wondered, must a pigeon be able to see from inside an aviary in order to be able to find its way home from a distance? To test this he raised another group of nestlings in the same aviaries but this time he surrounded them with a high fence which prevented the birds from seeing the horizon, though they could still see the sky. After a few months these also were released at Osnabrück. Not a single bird managed to find its way home. They had all been marked so that they could be identified if they were found anywhere else; many of them were in fact recovered, in places scattered in all directions from Osnabrück.

Clearly, homing pigeons need to be able to see the horizon at the place where they are reared if they are to be able to return home after being released elsewhere. Dr Matthews and others have suggested that they might do it by using their eyes in the same way as a navigator uses a sextant. At any particular point on the earth's surface when the sun rises the angle between it and the horizon continues to increase until noon, after which it decreases until the sun finally sets below the horizon. The speed at which the sun travels across the sky also varies with the time of day (see page 97). The suggestion is that a pigeon, using its internal clock, knows the altitude of the sun and the speed at which it is travelling at its home territory at any time of the day. If it is taken to any other place and released, both the angle and the speed will be different from those at home. The pigeon is able to work out from these differences which direction it must fly in order to return home. To be able to do this would require remarkable ability on the part of the bird, though not more remarkable than other abilities which we now know animals possess.

Unlike the eyes of other vertebrates the bird's eye possesses a mysterious structure known as the pecten. This is a frond-like outgrowth from the retina into the interior of the eye and shading that part of the retina known as the blind spot, which is the point where the optic nerve leaves the eyeball and where there are therefore no sensory cells. Its function has always been a complete mystery, but it has now been suggested that perhaps it acts as the sextant which Dr Matthews's theory postulates.

NAVIGATION BY THE STARS

So far we have concentrated on bird navigation by day, but many birds migrate at night, including the sea birds dealt with earlier in the chapter. How do they navigate when there is no sun to guide them? Much evidence is now available that birds are able to use the stars to navigate just as well as they are able to use the sun. And if the sky becomes clouded over, such night-flying migrants are lost just as those which rely upon the sun are lost if the sky becomes completely clouded over during the daytime. A dramatic demonstration of this was witnessed on 24 October 1963 in Münster, Germany. On that night the town was celebrating its traditional autumn fair, and the whole area was brightly lit, but the sky was completely overcast. Just after 7 o'clock the inhabitants attending the fair became aware of an intense noise of cawing birds overhead. Zoologists were able to make out a huge flock of between 300 and 400 cranes circling overhead, apparently attracted by the bright lights below. What had happened was that they had set out on their autumn migration, which they always undertake at night, and had been completely confused when dense cloud covered the sky. The lights of Münster became their only guide, and they were trapped in this light just as surely as a moth which comes into the house at night is trapped by the light in the centre of the room. For the next five hours or so the cranes circled above the city, flying in a clockwise direction, until after midnight, when the cloud began to thin and reveal patches of clear sky. The birds then flew away in a south-westerly direction along their traditional migration route.

The evidence which proves or at least strongly suggests

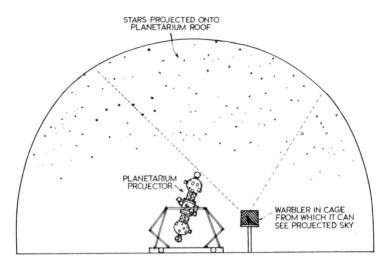

STARS PROJECTED ONTO
PLANETARIUM ROOF

PLANETARIUM
PROJECTOR

WARBLER IN CAGE
FROM WHICH IT CAN
SEE PROJECTED SKY

Dr Sauer's planetarium experiment by which he showed that blackcaps and
other small birds which migrate at night can recognise star patterns and
orientate in accordance with them

that birds which migrate at night use the stars for navigation
was produced in Germany by Dr Franz Sauer and his wife as
a result of their famous planetarium experiments in 1956.
The birds they used were blackcaps and other warblers
which spend the summer in Europe but migrate to Africa
for the winter. Their first observation was that, as the time
approached for them to migrate, warblers which they had
reared in cages sat on their perches at night facing the
correct migration direction, provided they could see the
night sky uncovered by cloud. This suggested to them that
the birds in fact used the stars to orientate themselves.

In order to verify their theory they arranged with the
Bremen planetarium to carry out tests with their caged
birds. The dome of the planetarium was first illuminated to
reproduce faithfully the star pattern as it was at the moment
in the real sky above Bremen. Each caged bird on its perch
showed no hesitation in facing south-east, the direction in
which others of its kind were even now flying to their winter
quarters. This was in itself remarkable, because these

experimental birds were fledglings of the year that the
Sauers had hand-reared away from all contact with their
parents or other warblers. The fact that they faced the
correct migration direction, however, did not actually prove
that they were using the star pattern in the planetarium to
orientate. They might be using the earth's magnetic field or
some other unknown force.

Further experiments, though, left no doubts. The cage
was covered while the whole planetarium sky was turned 90
anti-clockwise, so that the stars which should have been to
the south were now to the east. When the cover was removed
from the cage it took each bird but a few moments to
examine the new sky and orientate itself to the true
north-east, which of course was south-east according to the
star pattern. This experiment was repeated several times,
the star pattern being turned a different number of degrees
from true, and each time the caged bird faced south-east
according to the star pattern. When the stars were faded out
and the whole dome illuminated with a diffused light the
bird was completely confused and unable to fix upon any
direction.

Clearly, then, the blackcap about to set out from the
Bremen area on its autumn migration was able to choose the
correct direction of flight by observing the pattern of the
stars. But this was only the beginning. The route taken by
such migrating warblers was known to be south-east to
Turkey, where they then turned due south to carry them
across the eastern Mediterranean, through Egypt until they
reached their destination south of the Nile. In a further
series of planetarium experiments the Sauers were able to
show how the birds navigated over this very long journey, by
taking them through it stage by stage. Having first shown a
bird the stars as they are at Bremen, they then covered the
cage while the stars were rearranged to illustrate the star
pattern as it would appear above Prague. The bird when
uncovered almost immediately took up a position facing

south-east. The same thing happened when it was shown the night sky first over Budapest and then over Sofia. But when the sky pattern was changed to resemble the sky over the eastern Mediterranean in the region of Cyprus and Israel the bird suddenly changed its direction and faced due south in the direction of Egypt. Finally the bird was shown the sky as it would appear south of the Nile, its final destination. This time the restlessness with which it had greeted all the other star patterns it had been shown vanished, and instead of orientating itself on its perch it just went to sleep. Clearly it thought that it had reached the end of its arduous journey, though of course it was still in Bremen!

CHAPTER 8

NAVIGATION USING THE EARTH'S MAGNETIC FIELD

Forty years ago a suggestion that animals might be able to use the earth's magnetic field to navigate or orientate would have been dismissed as an impossible fantasy. But then so would suggestions that animals might be able to find their way around using sound waves or by producing an electrical field, by using the movement of the sun by day or the pattern of the stars by night. Today we know that various animals are able to use all these aids to navigation, so the possibility that they might be able to make use of the earth's magnetism no longer seems quite so fantastic. Investigations into the possibility that some animals at least might possess a magnetic sense are, however, still in their infancy.

The first concrete evidence resulted from a chance observation by Dr Hans Fromme at the Frankfurt Zoological Institute in the autumn of 1957. He had in a cage in the Institute several robins on which he was carrying out observations while, outside, robins were preparing to migrate south-westwards to Spain, where they normally spend the winter. Dr Fromme noticed that his robins were becoming restless, a not unusual phenomenon in migrating birds kept in cages at the time of migration, as we have already seen. But his robins were fluttering up only to the south-western side of the cage. Again not unusual. What was unusual was that the robins could see neither the sky by day nor the stars by night, since their cage was in a shuttered room. So any kind of celestial navigation could be ruled out. They must, Fromme argued, be responding to something

137

that they could 'feel' through the walls of the building. And what else could this be but the earth's magnetic field?

To test his improbable theory he put his robins' cage into a steel chamber. Outside this chamber the normal strength of the earth's magnetic field at Frankfurt is 0.41 Gauss, a Gauss being the unit in which magnetic force is measured. Inside the steel chamber the strength of the field was reduced to 0.14 Gauss. The effect of this reduction was very significant. The birds were still restless, because they wanted to be on their seasonal migration, but their flutterings were no longer directional. Clearly they no longer knew in which direction they should fly.

Other workers had already carried out experiments designed to find out whether birds had a magnetic sense, but these had produced no positive results. Homing pigeons, for example, had been fitted with small bar magnets, the idea being that if they used magnetic sense based on the earth's magnetic field to find their way home, these additional magnets would confuse them. But they had no effect on the pigeons' homing ability. Other birds were placed between two strong magnetic coils to see if this would confuse them, but again their directional flight was not affected.

Professor F. W. Merkel, a colleague of Dr Fromme's, now joined him, and suggested that if the strength of magnetic field was, as it seemed to be, important, perhaps if it was varied a sensitive bird might take some time to acclimatise to the new field strength. Accordingly the experiment with robins in the steel chamber was repeated, but this time the birds were left in the chamber for a number of days. Gradually they seemed to become adapted to the new magnetic conditions, for after a few days their flutterings became once more orientated, and they were again flying up to the south-west side of the cage. Coils were now brought into the chamber designed to change the direction of the magnetic field, just as Dr Sauer had changed the night sky in the planetarium (see Chapter 7). The results were similar. Given time to adjust to the new field, the robins now fluttered up in a direction which was south-west according to the new field. That they could use the earth's magnetic field explained why migrating robins could continue their migration flight when the sky became completely clouded over, whereas blackcaps and other night migrants had to land until the sky cleared.

Research is gradually showing that a variety of animals from widely different parts of the animal kingdom are sensitive to magnetic force. Towards the end of 1963 Professor Günther Becker of Berlin received a consignment of termite queens of various species from Rhodesia. As it was late in the day when they arrived he simply tipped them into a spacious box until the following morning. When he opened up the box he was astonished to find that they were all lying apparently asleep in an east-west position, and not in all directions as one would have expected. Intrigued by what he saw he carefully turned the box through 90 and replaced the lid. When he removed it again a few hours later he found that all the termites had changed their positions, and were again facing either due east or due west. He now put the box into a thick-walled steel chamber, which would exclude most of the earth's magnetic field. This time there was no order about the termites' arrangement when they were examined some hours later. They were lying in all directions. Finally he suspended a bar magnet above the box within the steel chamber. Within a few hours all the termites had orientated themselves so that they were lying across the axis of the magnet. It had been known for a long time that when the nests of many termite species are opened the queens are always found lying either in an east-west or in a north-south direction. What the significance of this orientation is remains a mystery.

One kind of Australian termite, commonly known as the compass termite, has always aroused considerable interest because its 'hills' are always built with the same orientation. They are large structures up to 13ft high and 10ft long, but only about 3ft wide. The long axis always runs due north and south. Presumably the termites used magnetic sense when constructing these relatively enormous homes. From the animals' point of view this particular alignment has considerable advantage. In the morning and again in the evening the long sides are warmed by the sun, but in the middle of the day, when the sun might be too hot, only the thin north edge is being heated, so that the 'hill' will absorb the minimum amount of heat.

Here we must leave the intriguing subject of magnetic sense. Research is still going on, and it is likely that in another decade we shall have as much evidence for its existence as we have for the other methods of orientation which are now well documented.

BIBLIOGRAPHY

BURTON, MAURICE. *The Sixth Sense of Animals* (London, 1973)

DROSCHER, VITUS B. *The Magic of the Senses* (London, 1971)

FRISCH, KARL VON *The Dancing Bees* (London, 1954)

LOCKLEY, R. M. *Animal Navigation* (London 1967)

MARSHALL, N. B. *The Life of Fishes* (London 1965)

MILNE, LORUS AND MARGERY. *The Senses of Animals and Men* (London 1955)

WILLIAMS, C. B. *Insect Migration* (London 1958)

INDEX

Agrostis infusa, 81
Albatross, Laysan, 124;
  wandering, 124
Alevin, 16, *15*
Allen, Dr G. H., 19
Alosa alosa, 24; *finta*, 25
Ammoecoete larva, 21
Ammoecoetes branchialis, 21
Anguilla restrata, 13; *vulgaris,*
  9
Ants, 110-1
Azimuth, 111, *102*

Bat, barbastelle, *59;* fish-
  eating, 59; horseshoe, 58,
  57; tomb, 60; vampire, 60
Becker, Professor Günther, 139
Bee, 103-9; dances, 106, *104*
Bison, 68
Blackcap, 134-6
Bullock, Professor T. H., 73
Burton, Dr Maurice, 70
Butterfly, monarch, 79-81,
  78; painted lady, 75-8

Callorhinus ursinus, 33
Caribou, 68
Chalcites cuccolus, 125
Chorizagrestis auxiliaris, 83
Clemens, Dr W. A., 18
Clupea pilchardus, 25
Coccinella bipunctata, 93;
  septempunctata, 93;
  undecimpunctata, 94

Cockroach, 99
Corvus cornix, 126
Cranes, 133
Cuckoo, bronze, 125;
  European, 127-8
Crow, Alberta, 126; European
  hooded, 126

Danaus plexippus, 78
Diomedia exulans, 124;
  immutabilis, 124
Dog, tracking ability, 70-2
Dolphins, 34, 65-7
Donaldson, Dr L. R., 19
Dragonflies, 90-2
Drosophila, 100

Eel, American, 13; European,
  9-13, *10*; silver, 9, 11
Electric catfish, 36, 46-7
Electric eel, 36, 44-6
Electric organs, 41, *42, 45, 46,*
  48, 49
Electrophorus electricus, 36
Elver, 13
Euphausia superba, 35
Eurygaster integriceps, 83

Fabre, Henri, 93
Firefly, 113-4
Frazer river, 18
Frisch, Professor Karl von,
  103-9
Fromme, Dr Hans, 137, 138

Galambos, Robert, 54
Glowworm, 113-4
Griffin, Donald, 53-4, 62
Grunion, 29-30
Gulf Stream, 13
Gymnarchus niloticus, 37-41, *39*
Gymnotus carapo, 42

Hartridge, H., 52
Hasler, A. D., 19
Herring, 22-4
Hippodamia convergens, 94
Humming bird, 124-5

Jurine, Charles, 51

Knife-fish, 41-3
Kramer, Dr Gustav, 128, 129, 130, 131, 132
Krill, 35

Ladybird, convergent, 94;
  eleven-spot, 94; seven-spot, 93; sixteen-spot, 95;
  two-spot, 93
Lamprey, sea, 21-2, *20*
Lateral line, 30, *31*
Lawrence, B., 65
Leptocephalus brevirostris, 10
Leptocephalus larva, 12-3, *10*
Leuresthes tenuis, 29
Lissmann, Dr H., 36-9
Lockley, R. M., 64, 116, 118
Locust, desert, 84-90;
  migratory, 84-90
Locusta danica, 87
Locusta migratoria, 84, 87;
  phase *gregaria*, 87, 89;
  phase *solitaria*, 87-9; phase
  transiens 88

Mackerel, 26-9
Malapterurus electricus, 25
Matthews, Dr G. V. T., 128, 132
Maurolicus pennanti, 28
McBride, Arthur, 65
Meganyctiphanes norvegica, 25
Merkel, Professor F. W., 138
Micraspis sedecimpunctata, 95
Möhres, F. P., 57
Mormyroid fish, 41-4
Mormyromast, 44, 46, *43*
Mosquito, 113
Moth, antennae, *112*; army
  cut-worm, 82-3; bugong,
  81-2; scenting ability 111-3;
  silver-Y, 89, *88*

Neuromast organ, 30-1

Oceanites oceanicus, 123
Oil birds, 62-3
Oliver, Professor, 94
Oncorhynchus gorbuscha, 16;
  nerka, 16; *tschawytscha*, 16

Papi, F., 101
Pardi, L., 101
Penguins, 67
Perdeck, Dr A. C., 128
Petrel, storm-, 64-5; Wilson's, 123-4
Petromyzon marinus, 20
Photinus, 114
Photuris, 114
Pierce, Professor, 54
Pigeon, homing, 131-3
Pilchard, 25-6
Pit organ, *72*, *73*
Planetarium, 134-6
Plover, American golden, 126;
  Mongolian, 126

Plusia gamma, 89
Pluvialis dominica, 126
Puffinus gravis, 123; *puffinus*,
  116; *tenuirostris*, 120

Rattlesnake, 72-4; pit organ,
  72, *73*
Redd, 14, 17
Reimann, Dr J., 110-1
Robin, 137-9
Romanes, G. J., 71
Rowan, Professor William, 126
Ruppell, Werner, 126

Salamanders, 69-70
Salmo salar, 16
Salmon, Atlantic, 15-7, *14*;
  fry, 16, *15*; king, 16;
  Pacific, 16-20; parr, 17, *16*;
  pink, 16; red, 16; smolt, 17
Sandhoppers, 100-2
Sanschi, F., 110
Sardine, 25-6
Sargasso Sea, 11, 13, *12*
Sauer, Dr Franz, 128, 134, 138
Schistocerca gregaria, 84
Schmidt, Dr Johannes, 11
Seal, Northern fur, 33-4
Seals, 67
Shad, allis, 24; twaite, 25

Shearwater, greater, 123;
  Manx, 64-5, 116-20; short-
  tailed, 120-1
Shevill, W. E., 65
Siskoff, Dr, 91
Skate, 47-8, *49*
Spallanzani, Lazaro, 51
Sphex aegyptius, 90
Stargazers, 47
Starling, 128-31
Sterna paradisaea, 121
Steven, Dr G. W., 26
Storks, 127
Sunn pest, 83-4
Swiftlets, 63-4
Szidat, Dr, 91

Termites, 139-40
Tern, Arctic, 121-2
Torpedo-ray, 47, *48*
Tragus, *59*
Twitty, Professor, 69

Uvarov, Dr B. P., 87

Vanessa cardui, 75

Wasp, solitary, 89-90
Whales, 34-5, 65-7
Williams, Dr C. B., 84-5, 95